WHO BUILT THAT?

WHO BUILT THAT?

The at-a-glance guide to the world's greatest buildings and their famous architects

ALAN PHILLIPS

CHARTWELL
BOOKS, INC.

A QUINTET BOOK

Published by Chartwell Books
A Division of Book Sales, Inc.
110 Enterprise Avenue
Secaucus, New Jersey 07094

This edition produced for sale in the U.S.A.,
its territories and dependencies only.

ISBN 1-55521-927-6

Creative Director: Richard Dewing
Designer: Roy White
Project Editor: Katie Preston
Editor: Lydia Darbyshire
Picture Researcher: Mirco de Cet

Typeset in Great Britain by
Central Southern Typesetters, Eastbourne
Manufactured in Hong Kong by
Regent Publishing Services Limited
Printed in Hong Kong by
Leefung-Asco Printers Limited

CONTENTS

This book is not intended to be a history of world architecture. It is, rather, a concise, chronological directory of architectural evolution, and consequently more will be said about the development of architectural styles and technologies than about the more abstract areas of theory and criticism, an approach that would necessarily require the author to judge one position against another.

However, as in all books that set out to illustrate the continuum of architectural development together with important periods of Revival and Renaissance, it is necessary to attempt to define architecture itself, and to define it in a way that will provide a structure by which the great buildings of ancient, classical, industrial and contemporary civilization may be compared.

Architecture may be defined as the practical art of meeting the requirements of providing shelter by enclosing space through the use of the science of construction and within a specific context.

Architecture is an art for two reasons. First, architecture requires great skill, which is the result of knowledge, intelligence and practice. Second, architecture has endowed itself with a duty to please, and it must, therefore, confront the complex subject of aesthetics, which requires all that is built to provoke the response of sensory perception. Architecture is practical because, unlike most of the other arts, its products are inhabited and, by being inhabited, must function properly in the service of human needs.

The most important aspect of the function of architecture is to provide shelter, be it for the god-kings of ancient Egypt, the audience of a concert hall or the modest family of late 20th-century civilization. As well as dealing with more prosaic issues such as climate, therefore, the practical art of architecture often requires the study of sociology, psychology and moral philosophy in making certain that the function of the shelter is appropriate to specific cultural, ethnic and behavioural phenomena.

COLOSSEUM, ROME

Space and form must be considered together, for one is always the consequence of the other. Architects have always been obsessed with the "will to hollow form". In doing so, space in which the function of human activity can take place is provided and in which the solid – walls, columns, beams, roof and floor – are the servants of the void. At this level, an understanding of space is straightforward. However, beyond the pragmatic comes a series of questions to which all the histories and theories of architecture provide no prescribed or universal answers. There are no rules in architecture that determine, for instance, how "high" the ceiling of a church should be nor how it should look, nor how "square" the foyer of a city hall should be. These are left to the will of the architect who may, nevertheless, be guided by the orthodoxy of a particular style or by the principles embodied in globally accepted theories of scale, proportion and composition, whereby space and form can be measured and judged by setting a build-

ing's size within the prescriptions of a pre-conceived mathematical system or order.

In tracing the evolution of architecture, this book will attempt to highlight key buildings whose aesthetic value is accepted because the arrangement of space and form adheres to the tenets of a particular architectural system or prescription. It will also consider those buildings whose appearance is so because they belong to the principles of a particular period or style, and those rare masterpieces that are never, by definition, typical but that are great because they are either progenitors of a new period or extraordinary for the principles they have disobeyed rather than those they have followed.

Science is fundamental to architecture. Even the most humble act of construction demands a knowledge of physics in reconcil-ing the forces of gravity, of chemistry in understanding the nature of materials and of biology in making buildings that harmonize with the growth and form of man and nature. The science of construction is concerned not only with engineering and technology as the means by which to prevent a building from collapsing but also with bringing the parts of a building together so that joints, junctions, surface texture, colour and materials combine in a state of balance and equilibrium to enhance the conceptual intentions of the overall form – and space-making activity. Within the relationship between the served space, function and the appropriate construction lies a moment of seamlessness between art and science.

The craft of architecture lies within the art of construction. A knowledge and mastery of

ST PETER'S, ROME

materials such as stone, timber, brick, steel and glass will enable the architect to fashion them into a constructed composition whereby the whole will always be greater than the sum of its parts. The same knowledge will give the architect the skills necessary to communicate his or her intentions to the stonemason, woodcarver or metalworker, so

that, if required, the ornament of architecture can reside within the building components rather than being applied afterwards.

All parts of the world have different climates, different geographies and different cultures. These conditions form the context of architecture. Climate may sometimes have a radical effect on architectural form, with buildings in hot countries being recognized, perhaps, by small openings, sunscreens and verandas, while wet or cold climates will encourage an architecture of large, sloping roofs, thick walls and shutters. However, this book will also show that in many major works and especially in contemporary buildings, which can rely on air conditioning, the importance of architectural form as the product of function and idea will enable the architect to ignore the exigencies of climate as a visual modifier. This is particularly true of religious architecture.

All buildings have sites, and all sites have surroundings. The immediate geography of an architectural site, be it low wetlands or a rocky hillside, will obviously influence the choice of structure – and, therefore, its form. However, the broader issues of context, which bear upon the degree by which an architect should take account of, say, an

DURHAM CATHEDRAL, ENGLAND

historical urban site or the beauty of a natural landscape, have raised more arguments concerning the appropriateness of an architectural statement than any other single issue. Some architects hold the view that they have a duty to respond to the spirit of their age – what is called the *Zeitgiest* – and that, whatever the context, a contemporary architecture has the potential simultaneously to respect history and to serve a creative continuum.

Other architects, however, believe that a new building in an historical setting should imitate the style of the original buildings, while others hold the view that classical architecture, for example, was so perfect, that all buildings in that style will always be in context because it is impossible for perfection to offend. Interestingly, this directory of world architecture will point to masterpieces that, whether they were built at the beginning of the 1st century BC or at the end of the 20th century AD, are nearly always radical, innovative and avant-garde, having little regard for climate or context.

Culture is another aspect of context. Countries have different languages, beliefs, religions, social habits and political persuasions, and attitudes to work, leisure, sleeping and eating vary, not only from country to country, but often from region to region. Poverty, wealth, trade and commerce influence activities, traditions and behavioural patterns. They all combine to establish a cultural milieu and regionalism that has architecture as its mirror.

Nevertheless, despite all these considerations that serve as an introduction to the complex subject of architecture, none is more important than the principle of idea. Behind all great buildings there is a great idea, which becomes the driving force of the architectural statement. It also remains the means by which the architect can judge his performance. When a project has been completed, the enabling or organizing idea behind it can be appraised and tested against the finished building. The architect and critic and history itself will have an opportunity to

GUGGENHEIM MUSEUM, NEW YORK

value the validity of the building to the idea that created it, and the architect can learn from mistakes or shortcomings and apply greater rigour to future projects.

Therefore, in addition to being a practical art, craft and science, architecture is, above all, an intellectual activity that has always been and will always be a fossil of civilization and a mirror of culture.

Who Built That? is a chronological record of the world's greatest buildings that starts at the beginning of the 3rd millenium BC and ends, 4,000 years later, at the end of the 20th century AD.

BLENHEIM PALACE, ENGLAND

STEP PYRAMID

SAKKARA, EGYPT IMHOTEP

The huge funerary complex of Pharaoh Zozer at Sakkara, Egypt, which was built between 2778 and 2760BC, was the first large monument to be made of stone. Many of the buildings were simply limestone versions of traditional timber and reed constructions, but the Step Pyramid in the centre of the complex represented a radical departure from tradition. This construction, approximately 65 metres (210 feet) high, was the first pyramid to be built. It was designed by the architect Imhotep (c.2800BC), who was worshipped as a god after his death.

The beginning of architecture – that is, building as a premeditated and conscious act of design – was extremely slow in coming. The first evolutionary stages of man's existence were tribal and nomadic. As hunter-gatherers, humans sought cover in caves and natural shelters, which were sufficient until the development of agriculture provided a reason for stability and permanence. Even so, for 5,000 years man seemed content with little more than merely utilitarian constructions as provided by locally available materials.

The earliest seeds of architecture would seem to have been sown in the Near East, where archaeologists have identified remains in Cyprus, Crete, northern Iraq and Jordan that would suggest a conscious effort to provide a planning arrangement that responded to function, and an interior construction that included ornament and an attention to culture and climate. This basic, rather rudimentary architecture took another 2,000 years to flower into the first period of Mesopotamian civilization, where the modesty of secular buildings gave way to the first great period of religiosity, but it is with a celebration of Egyptian building that our journey through architectural evolution begins.

PYRAMIDS AT GIZA

ANCIENT EGYPT

The 3rd of the 31 Dynasties of ancient Egyptian civilization was marked by the huge funerary complex that was built at Sakkara by the fertile banks of the River Nile for the Pharaoh Zozer, who reigned *c.*2667–2648BC, by the architect Imhotep. The first named architect in history, Imhotep, was responsible in his design for Zozer's necropolis for the first limestone ashlar walls, for the first engaged columns with carved stone capitals and for the first pyramid which rises in six huge steps and towers above the desert floor as a symbol of despotic power, god-king status and abstract art.

One hundred years later, the achievements of the Old Kingdom's monumental architecture achieved their culmination in the pyramids at Giza – the Great Pyramid built by the Pharaoh Cheops (Khufu), the second by Chephren (Khatre) and the third by Mycerinus (Menkaure). The austerity of form and power of the diagonal have inspired generations of architects, to no less an extent than the mathematical exactitude, with which the scheme was conceived and laid out, inspired them.

The Chephren group best illustrates the relatively static and dogmatic tradition of early Egyptian architecture in the beautiful formality of the arrangement of the pyramid itself – a huge cairn and a direct development of the earlier, sloping-sided mastaba tombs – the funerary temple on the pyramid terrace, the long causeway protected by the human-headed lion Sphinx and the valley temple on the edge of the Nile, which houses statues of the dead kings and was used for purification and embalming rites.

Many other pyramids were built during the 5th and 6th Dynasties of the Old King-

The pyramids of Giza, near Cairo, Egypt, were built between 2650 and 2563BC. The Great Pyramid was built for Cheops, while the other pyramids were added by Chephren (the son or brother of Cheops) and Mycerinus. The Great Pyramid is 230m (755ft) square and was once 147m (481ft) high – the top 3m (10ft) have worn away. These pyramids were built to last from blocks of limestone and granite. They were originally faced with dazzling white limestone, which must have had the effect of counteracting the rather ponderous monumentality we see today. The angle of the slopes of the Great

Pyramid was 52°, an angle that symbolized the slanting rays of the sun and that was intended to provide the soul of the king buried within a pathway to ascend to the sun-god.

HATSHEPSUT'S TOMB

DÊR EL-BAHARI, EGYPT **SENENMUT**

The elegant terraced mortuary temple of Queen Hatshepsut at Dêr El-Bahari was built by the queen's architect Senenmut *c.*1520BC. The terracing, colonnades and pillars that prefigure the Doric style give the temple a Greek gloss, but the causeways and sphinx-lined processional ways are in the Egyptian tradition. Hatshepsut's actual tomb is carved deep in the mountains behind the temple.

dom, and their subsequent decline was paralleled by the political and social upheavals during the dynasties that led to the dawn of the Middle Kingdom (*c.*2050–1786BC). The great Pharaoh Mentuhetep (Menthorpe) II of the 11th Dynasty reunified the territories and politics of Egypt and the new stability was personified in the building of the mortuary temple at Dêr El-Bahari in what has come to be known as the Valley of the Kings. A completely solid, and therefore

wholly symbolic, pyramid was raised above a huge colonnaded platform, with the rock-cut tomb driven deep into the sheer cliffs behind.

About 500 years later the first and only woman pharaoh, Hatshepsut (*c.* 1540–1481BC), declared her dominant personality not only by usurping the rights of her son, Tuthmosis II, but by building a far larger temple next to that of Mentuhetep II. The huge terraces and elegant colonnades were a return to the relative lightness of the architect Imhotep's funerary complex for Zozer, but the building of the temple broke with the seemingly timeless tradition of static monumentality that had prevailed for 1,500 years in the complex of valley temple, causeway and mortuary temple that was almost Grecian in character. The building was covered in coloured reliefs, which can now be read as a directory of the queen's favourite pursuits within the arts, the state and religion.

Spanning the Middle Kingdom, the New Kingdom (1567–1085BC) and the Late Dynastic Period (1085–332BC) and taking an extraordinary 1,200 years in construction is the Temple of Amon at Karnak, which was founded by Amenemhat (Ammenemes) I in the 12th Dynasty and continued by a succes-

Relief sculptures of the queen and the rich wall decorations depicting her triumphs feature heavily on the arches, colonnades and terraced walls of Hatshepsut's temple. She was the only female pharaoh to have ruled Egypt, and after her death her consort, Tuthmosis III (1504–1450BC), defaced many of her images.

TEMPLE OF AMON

KARNAK, EGYPT

An avenue of ram-headed sphinxes lined the avenue that led from the Luxor complex to the Temple of Amon at Karnak. Much of this great complex was built from sandstone, discovered at Silsilah near Thebes.

sion of pharaohs after the Hyksos invaders from Syria and Palestine were driven from the Nile delta at the beginning of the 18th Dynasty. Thothmes (Tuthmosis) I was the first to continue the building work at Karnak and was the first pharaoh to be buried in the corridor tombs of the kings cut out of the rock of the Theban mountains. Rameses I (1320–1318BC) commenced the great hyper-style hall at Karnak, which was continued with the additions of pylons, obelisks and liturgical architecture that evolved into a formal and axial plan of great sophistication. The Egyptian development of the column,

capital and clerestory window was synchronous with the continuing memorialization of symbols of power such as the sacred ram of Amon, which, with its lion's body, forms an avenue of sphinxes linking the Temple of Amon at Karnak with the temples at Luxor.

As Egyptian civilization flourished and subsequent pharaohs restored and reinforced the works of earlier rulers, it is interesting to note how the rate of progress in the evolution of culture varied in different parts of the world. While Egypt had built cities and temples, developed trade and industry and, through writing, had recorded a true

The great Temple of Amon (or Amun) at Karnak was the work of many kings. The original shrine was established c.2000BC, but building continued from 1530 until 323BC. Within the temple complex were smaller temples to Khons, the son of Amon, and the Pharaoh Rameses III (1198–1166BC). The court of Rameses' temple was lined with statues of the god Osiris, who gradually took over from Amon as the most important Egyptian god.

WHO BUILT THAT?

13

"civilized" state, visitors to Britain in the neolithic or new Stone Age *c.*2500BC were building no more than ditch-surrounded camps at a time when Imhotep had completed the funerary complex for Zozer. 700 years later, during which time some of the greatest monuments of Egyptian civilization were constructed, Britain had slowly evolved from the building of crude earthen burial mounds containing megalithic dolmen, to the Bronze Age artefacts of the Beaker Folk of Brittany. Within a few years of Rameses I having constructed the great hyperstyle hall of the Temple of Amon, the Bronze Age stone circles at Avebury (*c.*1800BC) and Stonehenge (*c.*1500BC) in Wiltshire were erected. However, notwithstanding the relative crudity of their construction and the lack of inscriptions, the geometry and arrangement of these stone circles according to a series of astronomical parallaxes relating to both solar and lunar calendars combined with axes aligned with the rising sun at the summer equinox of the northern hemisphere, can only presuppose a wholly different type of sophistication and the dawn of British civilization, albeit in the absence of government, god-kings, cities and trade.

Stonehenge at sunset. Constructed *c.*1500BC on Salisbury Plain in Wiltshire, UK, the stone circle is 32m (106ft) in diameter. Within the outer ring of 30 sandstone sarsens stood an inner circle of about 60 blue stones. Within that circle stood two concentric, horse-shoe shaped crescents, open at the northeast end. The blue stones were probably swept down to the Salisbury area by glacier. In the centre of the circle stands a flat sandstone slab, 5m (16ft) long, which was probably an altar. To the east of it is the hele stone, which casts a shadow on the altar as the sun rises on the day of the summer solstice.

The Acropolis was built at the command of Pericles during the golden age of Athens, the 5th century BC. Magnificently sited on top of a hill above the city, and in full daily view of Athenian traders in the *agora* (market place) below, it was a celebration of Athena, the goddess of wisdom and justice and the spirit and inspiration of Athens itself. The walled complex consisted mainly of temples: the Erechtheion, a temple built to replace an earlier one dedicated to Athena but destroyed by the Persians in 480BC; the Parthenon, the great shrine to Athena which once contained a huge statue in ivory and gold; and a small temple to Athena Nike, celebrating Athena as the bringer of victory to the city. The ruins of these buildings can still be seen, as can the Propylaea, or gateway, designed by Mnesicles but never finished because of the Peloponnesian Wars. Other buildings, most of which are now lost, include the Pinacotheca (picture gallery), the Theatre of Dionysus, the Odeion (music room) of Herodes

Atticus and the Stoa (sheltered colonnade) of Eumenes. There was also a gigantic bronze statue of Athena Promachos by Pheidias, which once dominated the skyline.

Just as the Acropolis dominates Athens, the Parthenon, the great temple to the goddess Athena, dominates the Acropolis. It was built between 447 and 438 BC by the architects Ictinus (Iktinos) and Callicrates (Kallicrates), with Phidias (Pheidias) as the master sculptor, and

has been described as "the most perfect doric temple ever built". Its visually perfect proportions are the result of masterly optical illusions. Long, horizontal elements that look straight actually curve upwards in the centre; regular looking columns taper towards the top and bulge slightly in the middle; and pillars designed to be seen against the open sky are broader than those that stand against a solid background.

ANCIENT GREECE

The links between Greece and Egypt developed first through Grecian military interventions during the 26th Dynasty (664–525BC) to rescue Egypt from the Assyrian yoke, and second by trade. Greek craftsmen and artists helped to restore, albeit for a relatively short time, the architectural characteristics of the Old Kingdom, until, in 525BC, Egypt fell under Persian rule for the first of two occasions during which a succession of pharaohs failed to establish the type of cultural, religious and political stability that enabled the great building works of earlier Dynasties to be undertaken. The last native pharaoh was Nectanebus II, but Egypt was effectively under Persian control until the arrival in 332BC of Alexander the Great, whose capital Alexandria became the intellectual heart of Greek scholars and artists. From the stepping stone of Crete in the Mediterranean to the Aegean Sea builders, artists, architects and craftsmen travelled by sea or through Asia Minor and Troy to play a part in one of the most important turning points in architecture.

Although the Greeks followed their predecessors in Egypt, Mesopotamia and Anatolia in establishing their principal architectural works in the service of religion, the revolutionary aspect was rooted in the belief that beauty must be considered as a subject in its own right and that the philosophical search for all things beautiful will naturally lead to the perfect temple. However, the Greeks, like many other early civilizations, were subject to invasions, warring and catastrophes before they settled to build the great monuments that mark the Hellenic and Hellenistic periods (650–30BC).

The highly developed culture that had flowered in the Aegean region had been all but devastated by the Achaeans *c.*1400BC and was completely destroyed by the northern Greek Dorian incursion of *c.*1100BC. Many Aegeans fled to Asia Minor, where, as Ionians, they reconstituted their culture and civili-

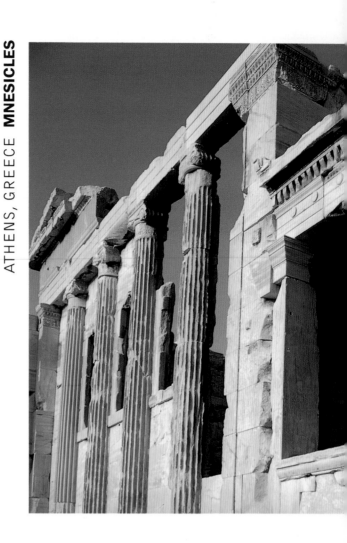

zation by building great cities. The Dorians began to colonize in the west, in southern Italy and Sicily. The Hellenic Greek civilization was born, although it was challenged by the Persians, who were finally beaten by the Greeks in the two great battles of Salamis (480BC) and Plataea (479BC), after which followed the great celebratory period of Greek building, which included the extraordinary Acropolis complex by the Athenian ruler Pericles (*c.*495–429BC). These temples derived from the Mycenaean megaron, or house of a chief, being a simple hall with a high, pitched roof and a front porch. Constructed entirely in timber, the earlier temples comprised a simple and elegant structure of columns, which supported beams over which was raised a simple pitched roof laid over longitudinal supports and without a truss. Painted terracotta tiles formed a protective frieze between the column heads and the eaves. Their study of the subject of

Built by Mnesicles between 421 and 405BC, the Erechtheion is a small, irregular, split-level temple built in the Ionic style. The architect had to fit the building into an awkward site, which was bounded by sacred groves and a burial ground. Unusual features include three porches, windows on each side of the door in the east porch and caryatids. These are the carved maidens used instead of columns to support the south porch. They lean slightly, as if emphasizing the weight of the roof they support.

aesthetics, combined with an innate sense of beauty and a relentless intellectual curiosity, which was not shared by the Egyptians, soon led the Greeks to consider that stone would be a more appropriate material than timber, although the new constructions so accurately imitated earlier timber detailing that the architecture was soon called "carpentry in marble". Scholarly inquiries into mathematics and geometry established the first principles of proportion and composition, and the precise ratios between all the parts of a column, from its capital to the stylobate or base platform, were defined.

An extraordinarily advanced sensitivity to form made Greek architects aware that optical illusions could be caused by the observation of linear vertical columns and uniformly horizontal stylobates. To correct the illusion of an inward curve that was created by the observation of a straight column, the shaft was tapered and made to bulge slightly. Similarly, a long, horizontal line of stylobates, architraves and cornices appeared to the Greek eye to sag at their middle. These, too, were modelled with a convexity in order to correct the optical illusion. Even the upper lines of lettering were exaggerated so that they appeared uniform when viewed from below. This art of aesthetic correction is called *Entasis*.

The two main branches of the Greek race, the Dorians and the Ionians, gave their names to the first two orders of architecture, the Doric and Ionic. The Etruscans developed the Tuscan order, while, after a lengthy period, the Hellenes devised the decorative Corinthian order, which matured under the hands of the Romans, who added the Composite order to establish the now famous five orders of architecture. Each order comprises an upright column, capital and base, and the horizontal entablature.

The Hellenistic period (323–30BC) was marked by Alexander the Great's expedition in 334BC, which subdued the Persian Empire, led to the submission of Egypt and supported his father, Philip II of Macedon, in establishing a peaceful Greece. On Alexander's death at Babylon in 323BC the animosity of the disparate Greek communities to the ad-hoc rule of his generals encouraged the centralized and united authority of Rome to intervene, and Greece became a Roman province, followed by Syria in 64BC and Egypt in 30BC. However, in spite of the political upheavals, the architecture flourished, and the second period of development moved away from religious building to concentrate on town planning, and civic projects, together with recreational and domestic buildings. The Romans were inspired by the Greek sophistication, although towards the end of the 1st century BC, taste declined as Roman constructional and technical virtuosity, together with a love of the most decorative and ornamental Corinthian order, interfered with the purity and simplicity of Grecian architecture.

The Pantheon in Rome, Italy, a temple to all the major Roman gods, was built in AD120–24 by Hadrian on the site of an old temple begun by Agrippa in 25BC. The rotunda reveals an impressive display of ingenious solutions to the problems of load-bearing inherent in such a building design. As round as it is high, 43.4m (142ft 6in), the rotunda rests on foundations that are 4.5m (14ft 9in) deep. The walls are 6m (20ft) thick, are made of brick-faced concrete, and are honeycombed with arches and buttressing, some of which forms niches for the statues of gods and emperors. Progressively lighter bricks were used as the walls rose, and the coffering in the ceiling was a decorative way of lightening the load. In the top of the dome an an oculus, or eye, 9m (30ft) wide, opens to the sky to provide the only light source and to make the dome less heavy.

THE ROMAN EMPIRE

After the fall of *Magna Graecia* and the decline of the Etruscan civilization, the Roman Empire emerged as a great republic able to take advantage of two colonies in which scientific, artistic and philosophical scholarship was without match.

During the early years of the Republic, the Romans' apathy to art and general conservatism makes it difficult to summarize their architectural achievements. After the fall of the Republic, however, the great generals, Sulla, Pompei and Julius Caesar, whose military victories provided a cause for monuments to be built in celebration, promulgated one of the great periods of Roman building. Caesar's heir, Octavian, or Augustus as he was later known, boasted that he had found Rome a city of bricks and left it a city of marble. In fact, the Roman development of the Etruscan arch into the vault and the ingenious invention of concrete as the principal load-bearing building material, would have Augustus' own successor, Tiberius (AD14–37), find a city of marble and leave a city of concrete.

The Augustan age, which dates from 27BC

and during which all emperors were given the surname Augustus, was one of the greatest eras in world history. The poets Virgil (70–19BC), Horace (65–8BC) and Ovid (43BC–AD17) recorded that rural life was unfashionable and populations flocked to the cities, which responded with massive building programmes under the inspiration and leadership of emperors such as Nero (reigned AD54–68), Hadrian (reigned 117–38), Caracalla (reigned 211–17) and Diocletian (reigned 284–305). These tireless patrons of architecture built thermae for games and bathing, circuses for races, ampitheatres for gladiatorial contests, theatres for dramas, basilicas for law suits and state temples for religion. During this period also appeared the apartment house or *domus*, which was organized like the empire in miniature, with the *patria potestas* or supreme power of the father, who protected family life with the same passion as the Augustan emperors sought to protect the numerous and far-reaching boundaries of their empire from barbarian attacks without any political discontent within.

While the Greek architects were artists, the Romans changed the role to that of technician and military engineer, although great works, such as the baths of Caracalla in Rome, were still under the hands of anonymous Greek sculptor–architects and craftsmen. However, the engineering legacy of the arch, vault and dome, combined with the Romans' virtuosity in concrete, stone and brick construction, was matched in scholarship by the writings of the Roman architect and engineer Marcus Vitruvius Pollio. Disappointed with the architectural styles of his Roman contemporaries, Vitruvius was concerned that the classical work of the Greeks should not be lost and his treatise, *De Architectura*, written in an appropriately Hellenist style, recorded the principles of symmetry, harmony and proportion. In 10 books, of which 50 copies survived into the Middle Ages, Vitruvius described the design of temples, theatres and civic buildings and included notes on construction, siting and the education of architects. From AD286, when the Roman Empire was divided into east and west with two separate emperors, and the eventual collapse of the Roman Empire in the west in 476 with the election of Odoacer as the first king of Italy, the 10 books of Vitruvius remained an invaluable bridge between the architecture of antiquity and its Renaissance in the 14th and 15th centuries.

The Colosseum, Rome, Italy, the largest Roman amphitheatre, was begun by the Emperor Vespasian in AD70 and completed by Domitian in AD82. A triumph of structural engineering, it included two concentric circulation passages, converging ramps of seating, stairways, vaulted ceilings in sloping tunnels and three tiers of arcading, with 80 arches on each storey. Lava, pumice, brick and tufa were used in its construction, and canvas awnings were drawn across the open top to protect the audience from rain or sun. The Colosseum could accommodate around 50,000 spectators, and it was still being used for games as late as AD523.

The Great Church of Holy Wisdom (Hagia Sophia) in what is now Istanbul, Turkey, was built for the Emperor Justinian by Isidore of Miletus and Anthemius of Tralles in the record time of five years between AD532 and 537. Inside, it shimmered with golden mosaics and its great dome, pierced by a ring of 40 windows, appeared to the contemporary historian Procopius to be "suspended by a chain from heaven". In fact it was supported on solid pendentives, each 18m (60ft) high. The dome supported by pendentives was the great Byzantine contribution to world architecture. When Constantinople fell to the Turks in 1453, Hagia Sophia became a mosque and minarets were added. It is now deconsecrated and houses a museum.

THE INFLUENCE OF CHRISTIANITY

The importance of architecture as the mirror of society was eclipsed in 313 when the converted Emperor Constantine issued the Edict of Milan, which gave Christianity equal rights with other religions and moved the capital of the Empire from Rome to the old Grecian colony of Byzantium, then re-named Constantinople and today called Istanbul. The early Christian period lasted from Constantine until the coronation of Charlemagne in 800, but not before many warring factions across Europe and Asia Minor brought chaos to the march of Christian progress.

Nevertheless, the collective enthusiasm and energy dedicated to the new religion, which caused more than 40 small churches to be built in Rome even before the Edict of Milan was promulgated, triumphed over the ravages of the Dark Ages and gave rise to the earliest post-pagan church architecture, which evolved, appropriately, from the Roman

St Mark's Cathedral in Venice, Italy, was built between 1042 and 1085, at the height of the powers of the Venetian Republic. Its Greek architect was inspired by the Byzantine Church of the Apostles in Constantinople, which was destroyed in 1483. St Mark's is set on the site of a previous church that had been built in 832 to house the body of the saint, smuggled from Alexandria by Venetian merchants. It was built on the Greek cross plan, with a large central dome and four flanking domes. The cathedral has been much bedizened since it was built: in addition to statuary, marble decoration and alabaster and porphyry columns, the gilded wooden domes were added in the 13th century, and the Gothic canopied niches and pinnacles were added in the 15th century.

basilica (from the Greek word *basileos* meaning kingly) or hall of justice to become the hall of God, the King of Kings. The basilican churches, of which there were 31 in Rome alone, were rectangular in plan, with twin colonnades separating a large nave from the side aisles, which were lowered to permit clerestory lighting above. The basilican church was generally erected over the burial place of the saint to whom it was dedicated, with the high altar above the crypt *confessio* or burial place.

In contrast to the great rectangular *basilikos* hall of congregational worship was the other type of Christian church, known now as Byzantine after the early Greek colony of Byzantium. Characterized by a centralized square plan covered with a domical construction supported on pendentives (concave structures), the Byzantine church, although found in the western Empire, established itself in the east as a symbol of Christianity in a divided empire after the death of Emperor Theodosius I in 395.

In 402 Honorius (395–423) removed himself from a malaria-ridden Rome to build in his adopted Italian city of Ravenna, which, by its geography, was heavily influenced by the Byzantine style. Justinian (527–65) reunited the Byzantine Empire during his reign, regaining the western provinces and marking the religious focus of his Empire with the masterpiece of Saint Sophia, the Great Church of the Holy Wisdom, in what is now Istanbul. In the centuries that followed Justinian, political divisions between east and west were matched by religious controversies. The eastern church maintained that the spirit proceeded from the Father only, while the western church held that the spirit proceeded from the Father and Son. The eastern Emperor Leo III (717–41), fearing idolatry, banned all representations of human and animal forms, and the absence of statutory and a preference for painted or mosaic ornament and religious representations are still the identifying mark of Byzantine architecture.

INDIA

While the Near East and Europe reeled under the religious schisms and political turmoil that led, through relentless attacks from the Muslims, Normans, Venetians and Ottoman Turks, to the decline of the Byzantine Empire, other civilizations were emerging. The earliest civilization to be recorded in India is the Indus Valley Culture, which developed slowly between 2500–1500BC. Military incursions then brought India into contact with many influences, including Graeco-Roman, Persian and Franco-British ones. Long-lasting indigenous empires and kingdoms rose and declined,

but no substantial architectural remains exist until the introduction of the Hindu, Buddhist and Jain cultures.

The Mauryan Buddhist Empire (321–184BC) left some rock-cut congregational shrines and *chaityas* or meeting halls. Post-and-beam supporting structures were of an Hellenistic type but were distorted so that they more directly represented a religion far removed from either Greek paganism or Roman Christianity. The religion of Jainism was founded by a contemporary of Guatama Buddha (*c*.563–483BC). Mahavira (599–527BC) was a brahmin who believed in salvation through successive rebirths. Jain temples

ORISSA, INDIA

LINGARAJA TEMPLE

Orissa in Northeast India was home to many temple-cities, each containing hundreds of Hindu shrines. One of the most important was the Lingaraja Temple at Bhubaneswar (Bhuvanesvara), which was built in AD1000. Just as the early Egyptians translated traditional wooden buildings into limestone, the architects of the Orissa temples copied ancient reed-built shrines. Reed shrines were built by bunching tall reeds together. The reeds were tied in a knot at the top, which was protected by an inverted bowl. This design can be seen, formally rendered in dry stone masonry, in the 55m (180ft) tall *shikhara*, or sanctuary tower, at Lingaraja.

TAJ MAHAL

AGRA, INDIA

The Taj Mahal, Agra, was built by Shah Jahan between 1632 and 1653 as a mausoleum for his favourite wife. It is a beautiful example of the Indo-Islamic style that characterized the Mogul period (16th to 18th centuries). The influence of Persian mosque building can be clearly seen – minarets, arched entrance and domes. The Taj is built entirely of marble and decorated by local craftsmen with a form of inlay work called *pietra dura*. Shah Jahan originally planned a matching tomb for himself to stand opposite the Taj, but ran out of money. He is buried with his wife in her mausoleum, both tombs enclosed by an exquisite white-pierced marble screen.

are similar to those of the conformist Hindu faith, differing only slightly in the amount of richness and complexity of sculptural ornament, which often obscures any structural purpose. Although the Hindu religion is the oldest, developing from faiths held by the Dravidian and Aryan invaders of the 15th century BC, modern Hinduism was established at approximately the same time as Christianity, with the belief that the Supreme Spirit exists in and works through the old Vedic Gods, Vishnu the preserver, Siva the destroyer and Brahma, the soul and creator of the universe. Thus, the brahmins, believing themselves to be directly descended from the Sanskrit-speaking Aryans, established a socio-religious aristocracy, which homogenized religion with politics in a way that affects every action and activity of Hindu life. Because man's way of life on earth is regarded as nothing more than a preparation for something more enduring after death, temples became more important than palaces, and the most complete examples were contemporaneous with the great examples of Byzantine architecture that marked the end of the first millennium.

SULEYMANIYE MOSQUE

ISTANBUL, TURKEY **SINAN**

THE INFLUENCE OF ISLAM

Muslim architecture also developed through the Indian subcontinent, but principally in part of the extraordinary influence of an Arabian Islamic Empire, which stretched from India in the east to Spain in the west.

While the architecture of the Roman Empire was characterized by the state, Muslim architecture is the product of a religion, especially of the teachings of Mohammed (*c*.571–632), whose words are recorded in the Koran and whose power was directed after his death by a succession of hereditary dynastic rulers, the Caliphs, to establish an architectural heritage that in quantity alone was a match for Mohammed's tireless enemy, the Byzantines. Early domestic and civic architecture does not survive, although Muslim life is engagingly documented in *The Arabian Nights* by Harun al-Rashid (673–809) and *The Rubáiyát*, which was written by the Persian poet, philosopher and mathematician Omar Khayyam (*c*.1050–1123). The Byzantine, or east Roman Empire, was itself overrun by the Seljuk Turks who succumbed to the Ottoman Turks in 1300.

The Suleymaniye Mosque, Istanbul, Turkey, was built between 1550 and 1557 by the great Ottoman architect Sinan. When the Turks took Constantinople (now Istanbul) in 1453, they brought with them the highly decorative Seljuk architectural style. Gradually the sober influence of the great Hagia Sophia was felt, and Ottoman architecture developed as an austere, undecorated, monumental discipline, concerned with producing the perfect, unified, domed space. The Suleymaniye Mosque owes its ground plan to Hagia Sophia. Its main dome is 55m (181ft) high and 25.5m (84ft) wide, and it is lit by 138 stained glass windows.

The Great Mosque at Damascus, Syria, is the earliest existing mosque in the world. Built by Caliph Walid between 706 and 715, it was erected on the site of a previous Christian church, which had itself been built on the grounds of an ancient Roman temple. The ruins of these old constructions were used in the building of the Mosque, which therefore incorporates some features more often associated with Christian architecture, such as aisles and a transept. The horseshoe pointed arches of its south arcade are, however, entirely Islamic.

The Alhambra, Granada, Spain was built mostly during the 13th and 14th centuries. The Moorish Palace within it was built by Yusuf I (1334–54) and Mohammed V (1344–91), the last rulers of the Moorish kingdom in Spain. From the outside, the Alhambra looks like a vast, dusty red fort (its name comes from the arabic *al-hamra*, "the red"); inside there are shaded, courtyards, tinkling fountains and pleasurable repose. Like all Muslim palaces, the plan lacks the formal symmetry of Northern European royal residences. It is based on two rectangular courts set at right angles and surrounded by a series of smaller halls and courtyards, baths, formal gardens, a small mosque and private apartments.

Both courts are open to the sky. The floor of the Court of Alberca is almost entirely taken up by a fish pond. The more elaborate Court of Lions measures 35 × 20 metres (115 × 66 ft), and surrounds a central fountain bowl upheld by the lions. Arcaded wooden *loggias* provide shade. Cool marble, slender columns, coloured enamel tiles, stalactite decoration (stucco arches moulded to look like natural forms) and the constant, refreshing presence of water in

fountains and artificial ponds are all typical of the "arabesque" style of architecture. The Alhambra was built for pleasure and recreation, and much of the moulding that would have been done in more permanent stone in a mosque, for instance, is here carried out in plaster. Even so, like most Moorish palaces, the Alhambra was intended to approximate to the Muslim idea of paradise: a temperate walled garden which is a haven from the desert heat.

THE FAR EAST

If an Arabian legacy to India and, particularly, Pakistan, is the Muslim mosque, then the Indian legacy to China must be the pagoda. Buddhism was introduced to China about AD65, and the pagoda was a strictly Buddhist building.

Before that, Confucius (c.550–478BC) had laid down a social, ethical and moral order that competed with the more mystical teachings of the Taoist teacher Lao-tze. Thus, three religio-socio-political systems dominated the early world of China, which foundered under a line of early despotic emperors, but was revived under the Han Dynasty (206BC–AD220), which so developed the economic and cultural state of the empire that, under Emperor Kuang Wu Ti, it challenged the Roman Empire of Hadrian as one of the greatest on earth. Chinese architecture, however, was never as dynamic or evolutionary as its cultural, civic and mili-

GREAT WALL OF CHINA

The Great Wall of China is 2250km (1400 miles) long, varies between 6 and 9 metres (20–30ft) in height and is 4.6m (15ft) wide at the top. The brick arches were added in AD1368. Originally a series of unconnected earth embankments and local defences, the wall was unified into one huge, impressive fortification in the 3rd century AD by Ch'in Shi Huang Ti, the first emperor of China. More than just a wall to keep out the northern nomadic hordes, it was an act of political will by an emperor who unified China, gave it a road network and imposed standard laws of literacy, currency and measurement.

SUMMER PALACE

BEIJING, CHINA

tary history. Although Buddhist influences gave China the multi-layered pagoda, the Chinese have, in general, had little time for the outward manifestations of religion, no territorial aristocracy and a creative disregard for domestic architecture. As a result there are few great temples, country houses or town mansions, and the few examples that have survived declare a national identity as much by the process of construction, and in

The Summer Palace, northwest of Beijing, China, was created during the Manchu Dynasty (1644–1911). A 333-hectare (832-acre) lakeside retreat, it is a masterpiece of landscaping, for

although the lake is natural, the hills, islands, woods and causeways are all artificial. Elegant marble bridges link the islands with the mainland, and there are over 100 buildings, most of which were erected in the 19th century. Most of the building is on the peninsula on the north of the lake. The main group of courts is dominated by the Fo Hsiang Ko, an octagonal tower.

particular the roof, than by an intellectual curiosity towards the development of the art of architecture.

It was another 450 years before Buddhism travelled from China, via Korea, to the volcanic shores of Japan, then a mysterious country whose prehistoric origins were unclear. However, the *Hojiki Record of Ancient Matters* (AD712), and the 8th-century chronicle *Nihougi* claims that there was an unbroken line of mikados or chieftains, starting with the country's unification under Emperor Jiminu in 660BC. Japan's immunity to external influences and self-imposed isolationism meant that a normal architectural chronology is difficult to identify. Instead, there has been a steady refinement of a centuries-old craft tradition that relies almost entirely on the indigenous materials of timber, clay, metal and fibre. In this respect, Japanese architecture is similar to that found

KATSURA PALACE

KYOTO, JAPAN **KOBORI ENSHU**

Katsura-rikyu, or Katsura Palace, Kyoto, Japan was built between 1615 and 1624 by Kobori Enshu, architect and master of the tea ceremony. Austere, intimate, tranquil and set in picturesque wooded gardens, the Katsura Palace is the antithesis of the ornate artificiality of western palaces being built at the same time. It is the quintessence of the *sukiya* style of architecture, embraced by the Japanese nobility from the 14th century onwards. *Sukiya* is characterized by cunning simplicity and

architecture harmonious with its natural setting – the interlocking of the house with the garden. In the Katsura palace, there are no murals or paintings; instead, the walls open to reveal tranquil natural landscapes.
The three main buildings that make up the palace are based on rustic vernacular buildings – a farm house, a mountain chalet and a tea-ceremony hut. The *Ko-shoin*, or entrance block, is the oldest part of the building. It contains guest rooms, a warming room and a

dining room. The *Chu-shoin* (middle block) contains small ceremonial rooms, some of which have *tokonoma* alcoves, where the only ornament is displayed. The *Shin-goten* (rear block) contains bedrooms, washrooms and a kitchen. Small tea houses and pavilions are placed about the gardens. The dimensions of all the buildings are based on multiples of the 1.8 × 0.9m (6 × 3ft) *tatami*, the traditional rice-straw mats used to cover the floor.
The Katsura Palace appears startlingly

modern to 20th-century eyes. Simple materials, often left in their natural state, the flexibility that sliding partition panels and doors confer, the use of standard prefabricated parts, modular planning (based on the *tatami*) and ornament that derives from geometry and construction features all predate the International style developed by the Bauhaus school.

MATSUMOTO CASTLE

In Japan the 16th century saw the building of great castles. These were constructed by the shoguns, or feudal overlords, and they were built, like medieval European castles, to both impress and defend. Most of them were set on stone or had granite walls, battered to resist the shocks of earthquakes, and were surrounded by a moat. Matsumoto Castle is in the prevailing style. The shuttered wooden galleries were designed to protect the shogun's army of archers while giving them excellent lines of sight.

in the Nile Valley in Egypt, with standard form-types changing little over 3,500 years.

The Japanese have also lived in the face of continuous natural disasters. Situated on the rim of the Pacific hurricane belt, with earthquakes and storms causing disasters on a huge scale, builders ignored the ample resources of stone to construct principally in timber. The impermanence of this material, coupled with a climate that brings heavy rainfalls and warm moist air from the Pacific in summer and a cold airstream from Asia during the winter, has led to a uniquely Japanese tradition of replacing important buildings with

exact replicas, sometimes as frequently as every 20 years. The result is a nation of facsimiles, with some Shiite shrines, for example, being contemporary rebuilds of monuments that were first erected 1,600 years previously.

After the prehistoric era came seven principal periods. The Asukak Period (538–645) saw the introduction of Buddhism, and close cultural and political ties with China were developed. Korean and Chinese artists and craftsmen settled in Yamato, bringing with them models and decorative techniques that created an almost instant art tradition that has survived to the present. The Nara Period (645–703), which was named after the new capital, developed a rectilinear city plan as a copy of the Chinese T'ang Dynasty capital Ch'ang-an. The Buddhist Emperor Shomn built temples in every province, of which the Shosm or Imperial Treasure House of $c.$AD752 still survives. The Helain Period in Japan (794–1185) coincides precisely with the advent of the Romanesque in Europe, which is marked by the coronation by the Pope of the Frankish King Charlemagne (800).

Durham Cathedral, County Durham, UK, an acknowledged masterpiece of Romanesque architecture, was begun in 1093 and finished in 1220. It was built by the monks of the great Benedictine abbey of which it was the church. Although the Norman apse at the east end was replaced between 1242 and 1289 by the huge Chapel of Nine Altars and the central tower was rebuilt after it had been struck by lightning in 1429, these later additions – usually in the Gothic style – do not detract from the cathedral's essential Romanesque character.

The Romanesque character of Durham Cathedral is seen to advantage inside the church. The round arches in the nave are supported by massive, decorated circular piers, which alternate with clustered piers. The ribbed vaulting, which is carried throughout the nave and aisles, is the earliest known example of the style.

MEDIEVAL EUROPE

After the chaos of the Dark Ages that followed the fall of the Roman Empire, states and nations began to evolve. At the same time science, letters, arts and culture were developed by the monastic civilization that exercised a Christian order over an emerging feudal system, in the same way as the Samurai developed almost total military power over an identical system in Japan. But whereas the Japanese declined to lose its scholars to China and Korea by reverting to a complete and deliberate isolation that mummified any architectural development, the nations that were emerging in western Europe were becoming powerful enough to set aside the rule of the Holy Roman Empire, and they no longer stood in awe of the architectural remains that memorialized the genius of Roman architecture. Thus, Romanesque was born from the ruins of ancient buildings and developed as a prologue to the great European period of Gothic architecture. Romanesque was a compound of many influences, including Roman, Byzantine, Carolingian and Ottonian, Viking Celtic and Saracenic, but it was universally influenced by the monastic churches of St Bernard and St Benedict, where an eclectic style was homogenized by the development of the Roman vault.

The church and castle were the two great building types of the early Middle Ages, and both suited the sober and dignified style of Romanesque architecture. Although 18th- and 19th-century critics considered this period minor in the evolution of world architecture, the rigour, experimentation and tireless enthusiasm of Romanesque builders, who were regarded as being on a par with harness-makers, cobblers and pot-makers, established an architectonic *Zeitgeist* without which the history of Gothic architecture could not have been written.

The painter and architect Giorgio Vasari (1511–74) is credited with the invention of the word Gothic as a disparaging term for an

CAERNARVON CASTLE

GWYNEDD, WALES

Caernarvon Castle, Gwynedd, Wales, was begun by Edward I of England (reigned 1272–1307), and it is typical of the many castles built in Wales or on the Borders in order to subdue the Welsh. Much had been learnt about military architecture from the Crusades, and these castles were designed so that a small force of trained men could hold off much larger numbers of attackers. Caernarvon Castle, which was built between 1283 and 1328, has a single outer wall, punctuated with polygonal towers. These were arranged to give cross-fire at every point and thus take advantage of the great improvement in long-bow and cross-bow technology.

Canterbury Cathedral, Kent, UK, is a re-modelling of a Norman church that had been built between 1071 and 1077. The result is a Norman–Gothic hybrid. Some fine Norman features remain, including a timber tower and a groined vault. After extensive rebuilding and fire damage, the eastern end was finally rebuilt between 1174 and 1185 by the French master-mason William of Sens.

CANTERBURY CATHEDRAL

KENT, UK **WILLIAM OF SENS**

MILAN CATHEDRAL

Milan Cathedral, the largest medieval church in Italy, was largely built between 1385 and 1485, but it was not entirely completed until the 19th century, when Napoleon finished off the façade. The intricate, soaring pinnacles, the delicate, traceried windows and the flying buttresses are the work of more than 50 architects and sculptors, most of whom came from the north, bringing their Gothic expertise with them. They made Milan a unique example of northern Gothic.

architectural style that was considered base for having departed so radically from Graeco-Roman classicism. But while Vasari's time was an age of Romanesque and Revival, the Gothic period, which lasted for four centuries throughout Europe, was the Age of Invention.

The early Gothic architects, or master-masons, began by examining the problems of constructing the Roman cross-vault. Because these were heavy and difficult to construct, the masons began instead to construct a skeleton of stone and mortar ribs, within which were laid the stone panels of the vault itself. Subsequently, a geometrical inversion took place. In the Roman vault the groin – that is, the line that occurs at the intersection of two vaults – occurred as the by-product of the shape of the vault. The Gothic builders started by laying out and constructing the groins, which determined the geometry of the vault. In this way the ribbed vault was developed, as was the pointed arch, which began in Mesopotamia and was passed to Sassanian Persia, whence it was used by the conquering Muslims across the whole of their empire.

The ribbed vault heralded a new structural economy through the application of the

principles of counterpoise. The collective weight of high vaulting, which bore both downward and oblique loads, was resisted and brought into equilibrium by a combination of dead-weight arches, which pressed against the clerestory nave walls, and huge buttresses and flying buttresses, which were weighted with ornamental pinnacles to bring the loads to ground. The pointed arch was both aesthetic and a means of resolving the problems of vaulting over an oblong area. However, a ceaseless striving to build vertically, which culminated in the symbolic spire of the Gothic cathedral, was wholly aesthetic and devotional in inspiration.

The baptistry, which forms part of the ensemble of cathedral, campanile and baptistry at Pisa, Italy, is a typical example of the Italian version of the Romanesque. External decoration was a special feature, although some of the arcading is a 14th-century Gothic addition. The baptistry was designed by Dioni Salvi and built between 1153 and 1278. Separate baptistries were a feature of 13th-century ecclesiastical architecture. Baptism was confined to three times a year – Epiphany, Easter and Pentecost – and large buildings were necessary to house the multiple ceremonies.

NOTRE DAME CATHEDRAL

The heightened status accorded to Gothic building brought with it a new respectability and sometimes renown for the architects of medieval Europe, and as their positions improved, the architects – still known as master-masons – came to concentrate on design and supervision, sometimes travelling great distances from project to project, and exchanging ideas, inventions and craft secrets through the establishment of guilds and lodges. Within these lodges also developed the arts of sculpture and stained glass, the chief servants of architecture. The new structural system of framing provided the opportunity for huge windows, or walls of glass, in which could be set massive pictorial compositions of coloured glass to depict religious events to a predominately illiterate congregation. As towns grew in size and prosperity, they became increasingly independent of the power of monastic dynasties, which were less interested in art, science and scholastic rationalism than in mysticism and contemplation. Nevertheless, French Gothic architecture was introduced to Britain by the Cistercians, and the architect

William of Sens was invited to rebuild the east part of Canterbury Cathedral. William of Sens's fame was matched only by that of Villard de Honnecourt of France, Arnolfo di Cambio (1232–1301) of Italy, Peter Parker (1333–99) of Germany and the English architects Henry Yevele (1320–1400) and Hugh Herland (1360–1405), a master-carpenter.

France remains the birthplace of the Gothic style. Milan Cathedral stands as a rare exception to Italy's reluctance to accept the Gothic over a developing Romanesque. Spain, too, built in the Gothic style, although the spirit was dampened by continuing political disunity, which prevailed until 1470, and the late arrival of the Gothic style in Germany resulted in originality being shown only in the *Hallenkirchen* or hall-churches.

The greatest celebration of late medieval architecture was seen in England, and the early English period, which lasted until the Black Death of 1348–9, was followed by the Decorated period of *c.*1250–1370 and by the long Perpendicular period, 1330–1540. Apart from the great building works in the cathe-

The Cathedral of Notre Dame in Paris was built between 1163 and 1250, and it is the prototype French Gothic cathedral, becoming a template for future cathedral design. A wide nave, double aisles, stubby transept and a chevet (circular or polygonal apse) formed a typical groundplan. The soaring height of the nave (33m/110ft), the pointed arches and the flying buttresses characterize the Gothic style.

The Cathedral at Chartres, France, was built on the site of an older Romanesque church that had burned down. Its foundations and crypt were used as a basis for the cathedral, which was begun in 1194 and "topped out" (roofed) in 1220. It was finally consecrated in 1260. Built between 1150 and 1300, a time when cathedrals were being raised all over France, Chartres has Gothic features without being such a paradigm of the style as Notre Dame in Paris. The nave at Chartres is vertiginously high – 36.5m (120ft) – and this gave the architect room to include more windows than usual, setting the Gothic fashion for more window than masonry in an expanse of wall. Unique features include the three triple-doored porches on the north and south sides and west front, and the non-matching towers on the

west front, which exemplify the difference between early and late Gothic. The austere octagonal pinnacle of the south tower is 12th-century, early Gothic; the pinnacle- and buttress-encrusted late Gothic north tower was added c.1507. Chartres is most renowned for its stained glass and sculpture. Beautiful narrative sculptures enliven the

doorways of all three porches. The 176 stained glass windows, described by John Ruskin as "flaming jewellery", are the finest example of 13th-century glass in the world.

Chartres was built in the town centre, a cathedral for the people rather than an adjunct to a monastery. The people of Chartres raised money for the

building by public subscription and, according to legend, themselves pulled the carts carrying the stone for the building from the quarry. The sculptures and stained glass pictures, which tell the story of the gospels and explain points of doctrine, were created for an instructive as well as aesthetic purpose, to help the people in their worship.

SALISBURY CATHEDRAL

WILTSHIRE, UK

drals of Exeter, Ely, Wells, Salisbury and Lincoln, the natural abundance of limestone and oak combined with the wealth of the church to make possible the construction of more than 9,000 parish churches in a Gothic vernacular that celebrates the end of the Middle Ages with a display of timber crafts-manship that has no parallel, except in the palatial architecture of China and Japan, where a timber craft traditional has prevailed in an unbroken line for over 1,500 years.

Salisbury Cathedral, Wiltshire, was the only English cathedral to be built on what might be called a "green field site", uncluttered by earlier church foundations. It was built in its entirety in only 40 years, between 1220 and 1260. (Renovations by Sir George Gilbert Scott (1821–78) took place in the 19th century.) Salisbury's uncluttered elegance of line and simple decoration make it the jewel of the early English Gothic style. English cathedrals built in this period went for length rather than height – Salisbury's nave is a mere 25.6m (84ft) high – and the double transept is unique to the English style. Seen from afar, Salisbury's most impressive feature is the spire, 123m (404ft) tall, once the tallest in the medieval world. According to structural engineering theory, it should not be able to stand; just as, according to aerodynamic theory, bumblebees should not be able to fly.

FOUNDLING HOSPITAL

FLORENCE, ITALY **FILIPPO BRUNELLESCHI**

The Foundling Hospital in Florence was the first major building undertaken by Filippo Brunelleschi, the architect who built the dome of Florence Cathedral, the first sign of the emergence of a coherent Renaissance architecture. The hospital, which was built between 1421 and 1445, was the first building Brunelleschi made on his return from Rome, where he had been studying classical architecture at first-hand. Brunelleschi was the first architect to realize that the language of ancient Roman architecture could be transferred to modern buildings. The loggia for the hospital has an arcade of Corinthian columns supporting round arches, with terracotta medallions in the spandrels. This Romanesque arcade was much copied by later Renaissance architects, the best example probably being the arcaded court in the ducal palace at Urbino, which was built *c.*1468 by Luciano Laurana.

THE RENAISSANCE

The 14th century saw a new spirit of intellectual inquiry, which was democratized by the invention of printing and radicalized by the Reformation in religion and a Renaissance, first in literature and, afterwards, in architecture. In England the new wealth of the laity led to the development of domestic architecture, which was encouraged by Henry VIII's dissolution of the monasteries. In Italy, the seat of Roman antiquity, the scholarship of Dante Alighieri (1265–1321), Petrarch (1304–74) and Boccaccio (1313–75) was celebrated by a national interest in classic literature, which paved the way for a revolt against medieval art.

The fall of Constantinople to the Turks in 1453 led to an influx of Greek scholars to Italy, especially to Florence, a city of great military and political power. Florence had conquered Pisa in 1406 to gain a valuable seaport, and had gone on to take Milan and Lucca, becoming a powerful republic, the centre of European Renaissance art and a place of religious intrigue, from where the pious and indefatigable Dominican friar

Savonarola (1452–98) simultaneously challenged the papal authority of Alexander VI and the tyranny of the powerful Medicis. Rome meanwhile awoke from the poverty of medieval feudalism to welcome the return of the popes from Avignon. Nicholas V (1447–55), Julius II (1503–13) and LeoX (1513–22) became great patrons of the arts.

The restoration of Roman socio-religious culture had, however, little influence over the Venetian state, whose great wealth from oriental trade, albeit dented by a more or less continuous war with the Turks after the sack of Constantinople, made Venice unsympathetic to spiritual control. Its rulers preferred to remain semi-independent of the popes and protective of a unique commercial system, defended by an island geography and powerful navy. The Venetian printing presses of John of Speyer, which were established in 1469, and the Aldine Press with its editions of Greek classics, transformed intellectual life by providing hitherto unobtainable information, including the first

The Louvre in Paris, now an art gallery and museum, was once a royal palace. Like many French palaces, it grew by accretion over the centuries, and shows elements of medieval, Renaissance, Baroque and neoclassical styles. In 1527, François I commissioned the remodelling of the existing medieval chateau. Work started under Pierre Lescot in 1546, but most of it was overseen by Henri II after his father died in 1547. Henri's own death in 1559 left the building unfinished. In 1627, Jacques Lemercier (c.1585–1654) added the Pavilion d'Horloge, and in 1667, Louis XIV recruited the great Bernini to finish the building. Bernini's plans were never used. Instead, Le Vau (1612–1670), Lebrun and Claude Perrault (1613–1688) provided the east front known as the colonnade. In the early years of the nineteenth century, the Louvre was, inevitably, improved by Napoleon. In 1852, Visconti and Lefuel provided ornate neorenaissance additions. Therefore, when I. M. Pei was commissioned to build an extension in 1989 in the Cour Napoléon, he chose a pyramid as the only possible architectural shape that could both complement the many existing styles and at the same time make its own unique statement.

CHURCH AT WIES

The pilgrim's church at Wies in Bavaria was built by Dominikus Zimmerman between 1746 and 1754. It is the most glittering example of the churches in southern Germany built by German architects heavily influenced by the Italian style. The interiors were particularly ornate in the Baroque sculptural style, and were characterized by gilding and bright, clear colours.

SAN CARLINO

ROME, ITALY **FRANCESCO BORROMINI**

Built in two stages by Francesco Borromini, San Carlo alle Quattro Fontane ("San Carlino") is a compact Baroque gem – a distillation of the architect's style. The main body of the church was built between 1638 and 1641. It is crammed onto a small corner site and was built on a plan of intersecting ovals reminiscent of Michelangelo's original, and unused, plan for the mighty St Peter's. This made it possible for the façade, added by the architect in 1665–1667, to follow the curving, sinuous line that was to characterize Borromini's work from then on. Inside the church, the space is so small it is said to fit into a single pier of St Peter's. The interior architecture is restless, complex, sinuous and hard to read in a gloom scarcely pierced by the single light source. Borromini himself was a complex, dark character and he committed suicide in 1667.

FLORENCE CATHEDRAL

ITALY **ARNOLFO DI CAMBIO, FILIPPO BRUNELLESCHI**

biography of the great architect Filippo Brunelleschi (1377–1446). In this biography, which was titled *Life* and was published *c.*1486, the anonymous author demanded that everyone should follow Brunelleschi's example of a new architecture that sought to recover and restore the "good, ancient manner of building".

The trans-Italian spirit of revival was symbolized by the literati's adoption of Ciceronian Latin, and the spirit of adventure was marked in 1334 by the painter Giotto di Bondone's appointment as architect to Florence in recognition of a new understanding of the status and position of architecture as an holistic art above and beyond the medieval notion of a master-mason. Giotto's appointment also signalled the beginning of the cult of the individual. Hitherto, medieval masons and carpenters,

Florence Cathedral, Italy, which is also known as Santa Maria del Fiore, was begun in 1296 by Arnolfo di Cambio and finished in 1461. The long building period meant that the main body of the cathedral is in the low-slung, non-buttressed Italian Gothic style, while the dome is a blend of Gothic and Renaissance styles. The huge pointed dome, 42m (138ft 6in) in diameter, was the work of Brunelleschi, who began it in 1420. He used techniques taken from the architects of Imperial Rome, including the herring-bone brickwork for extra strength, and combined them with such Gothic features as slender ribbing and a double-shell construction within the dome. He also installed the first timber and iron hoop tie, which was installed around the base of the dome to prevent it bursting outwards and to obviate the use of buttressing.

SANTA MARIA NOVELLA

FLORENCE, ITALY **LEON BATTISTA ALBERTI**

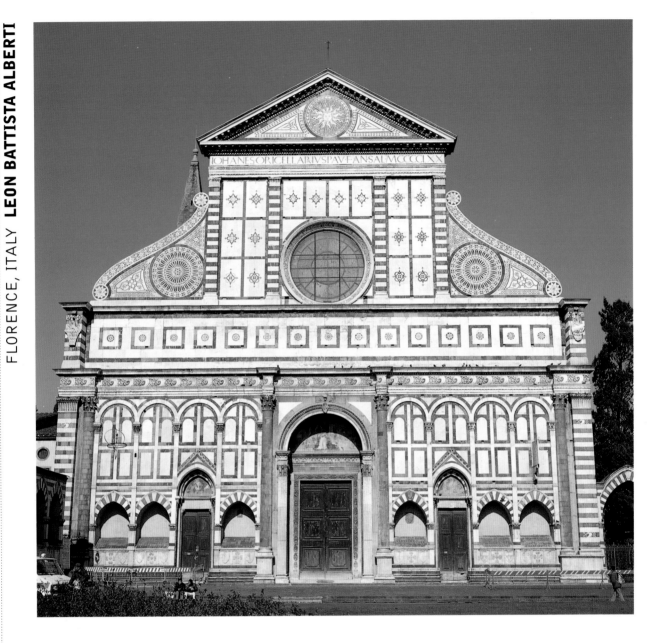

The harmonious façade of Santa Maria Novella in Florence, Italy, was designed by Alberti between 1456 and 1470. Its proportions, the coloured marble decoration and the scrolls that unite the nave and the aisles were much copied in later Renaissance churches. The original Gothic church was built between 1278 and 1350.

although known by name and sometimes renowned for great craft skills, were never credited with the intellectual notion of a preconceived and theoretical act of design. Although Chartres Cathedral in France and Salisbury Cathedral in England are masterpieces of structure, construction, craft and ornament, they were never understood as the product of a single genius.

If 1334 marked the elevation of architecture to the status of the "Mother of Arts", 1377 celebrated the birth of Brunelleschi, the first Renaissance architect who, in common with all the great masters that followed, was also skilled in sculpture, poetry, painting, engineering and mathematics. Brunelleschi

trained as a goldsmith, moved to architecture and engineering and combined his skills with the discipline of mathematics to establish the first understanding of perspective through the use of a vanishing point. The author of *Life* tells us that Brunelleschi: "perceived the method and the symmetry of the building technique of the Ancients, and he thought he could discern a certain order, as it were of bones and members, and it was made clear to him, almost as if God had illumined him. He sketched almost every building in Rome and the surrounding countryside, with measurements . . . as closely as he could estimate them."

In these words, the idea of the Renaissance

SANTA MARIA DELLA SALUTE

VENICE, ITALY **BALDASSARE LONGHENA**

The octagonal church of Santa Maria della Salute is the masterwork of Baldassare Longhena. Built on the Grand Canal, Venice, Italy, between 1631 and 1682, it is a supreme example of Venetian Baroque. The great scrolled buttresses, or volutes, which support the main dome are a particular Baroque feature. Venetian architecture always looked to the east as well as to the west, and Byzantine influence may be seen in the large, domed, central space.

ambition becomes clearer. Scholars were searching for mathematically harmonious proportions, as illuminated by the study of nature and the human form, symmetry, as found in the geometry of Greek architecture, and the sublime resolution of light, dark, solid and void towards an architectural composition of clarity, harmony and repose. The Italian humanist and architect Leon Battista Alberti (1404–72) continued the new age of scholarship with theoretical works on painting, *Della Pittura*, sculpture *De Statua* and the education and ethics of family life, *Della Famiglia*. Alberti's contemporaries, Brunelleschi, Donatello (*c.*1386–1466), Lorenzo Ghiberti (*c.*1378–1455), Masaccio (1401–*c.*1428) and Luca della Robbia (1400–82), provided a creative milieu in Florence that encouraged Alberti to unravel the complex analyses of Vitruvius and to produce, again in Latin, the influential treatise *De Re Aedificatoria*. When this work was published

VILLA CAPRA

VICENZA, ITALY ANDREA PALLADIO

Palladio's Villa Capra (also known as the Rotonda) just outside Vicenza, Italy, was designed as a "weekend retreat" from the vicissitudes of town life. It was begun in 1567, using modest materials – plastered brick, with stone ornaments. The Rotonda's severely classical, symmetrical plan, with porticos on all four sides to give views from every part of the house, was extremely influential and much copied in Europe and the United States.

in 1485 it synthesized the principles of scale, proportion and composition and established the most influential canon of Renaissance architecture, providing the basis of an architectural language comprising a vocabulary of base, column, pediment, entablature, arch and dome. In addition, multi-level buildings with, for example, shops at a lower level, living accommodation on the floor above and storage or servants quarters above, established a tripartite division of buildings. In this style the base or basement was often of heavy, rusticated stonework, the *piano nobile* or "level of greatest importance", being in the centre, had tall windows to illustrate the relative importance of the function behind, while the upper or attic level was marked with small windows, to indicate that the hierarchy is reversed to declare a triumph of solid over void.

This language was inherited by Donato Bramante (1444–1514), whose friends and

mentors included Leonardo da Vinci (1459–1519), Alberti and Piero della Francesca (c.1420–1492). Within this extraordinary environment, Bramante, who had trained as a painter, studied the work of Brunelleschi and turned his genius to architecture. He collaborated with Leonardo da Vinci in the Santa Maria delle Grazie, a partnership that gave Milan a great building and the *Last Supper*. The French invasion of northern

Italy forced Bramante to flee to Rome, where he taught Raphael (1483–1520) and the influential architect Antonio da Sangallo the Younger (1483–1546), and he was commissioned to design the new St Peter's by Pope Julius II. After Bramante's death in 1514 and the sack of Rome in 1527, Michelangelo (1475–1564) inherited the task of continuing the project, which was to become the apogee of classical architecture.

In common with Raphael, Michelangelo declared himself dissatisfied with the strict austerity of Bramante's work. With a classical language of infinite variety and possibility at their disposal, Michelangelo, Raphael and his pupil, Giulio Romano (1492/9–1546), simultaneously embraced the rigorous disciplines of a prescriptive and pedagogic high Renaissance while finding a more personal and indulgent expression in a variation of

FONTAINEBLEAU

FRANCE GILLES LE BRETON

The Chateau of Fontainebleau is a splendid expression of the French Renaissance style, which largely depended on the construction skills of French master-masons and the decorative creativity of Italian artists. Fontainebleau was built between 1528 and 1540 by Gilles le Breton for the flamboyant king François I. It started life as a medieval castle-cum-hunting lodge, which is why its plan is so irregular. Its most famous feature is the Galerie François I, a long gallery connecting the old castle keep to the front of the building. This was the first instance of such a gallery, which was later copied in many European countries. The moulded stucco decoration, alternating with painted panels, was the first of its kind seen in France. The artists were Italian, Giovanni Battista Rosso and Francesco Primaticcio. This ornate, Mannerist style became known as School of Fontainebleau, and was extremely influential.

the classical style that became known as Mannerism. An attitude to architecture, rather than a style, Mannerism is the product of a theoretical position that extended the austerity of a revived classicism, dedicated only to the ideals of harmony and unity through the mathematical arrangement of an antique vocabulary, to make a richer, more sensuous architecture that was capable of capriciously distorting or inverting the principles that had formerly governed the design of Renaissance buildings, partly in the name of aesthetics.

Mannerism did not, however, signal the arrival of a great period of Baroque architecture. The north Italian and hitherto parochial oligarchical republic of Venice, which had,

In the 16th century Antwerp, Belgium, was the richest city in northern Europe, and its magnificent Town Hall is appropriately grand. Built in 1561–6 by Belgium's foremost architect at the time, Cornelis Floris (or de Vriendt, 1514–75), it demonstrates the influence of the French rather than the Italian Renaissance. The large windows are typical of the northern Renaissance style. The stepped centre piece, with its classical decoration, interrupts an otherwise restrained façade. The stepped gable is characteristic of architecture in the Low Countries.

nevertheless, been home to Giovanni Bellini (*c*.1430–1516) Giorgione (*c*.1477–1510), Titian (*c*.1487–1576), Veronese (*c*.1528–88) and Jacopo Tintoretto (1518–94), provided a successor to Alberti in the great architect Andrea di Pietro della Gondala, now known as Palladio (1518–80). A serious student of the classics, and of Vitruvius and Roman architecture in particular, Palladio wrote his own opus, the *Quattro Libri di Architectura*. Published 10 years before his death, these four books demonstrated a profound understanding of classical systems of proportions and composition, an understanding that had enabled him, especially in domestic work around Vicenza, to manipulate these academic and scholastic foundations into a brilliant architecture that was not a moribund facsimile of history but hailed the past only in order to celebrate the future.

Palladio's pupil, Vincenzo Scamozzi

(1552–1616), used his book *Idea de l'Architectura Universale* (1615) to carry his and Palladio's ideas into the 17th century, and these important theoretical works, together with Vitruvius's original 1st-century writings, Alberti's *De Re Aedificatoria* and the documentary *Le Vite de'più eccelenti Architetti, Pittori, et Scultori Italiani (Lives of the Painters, Sculptors and Architects)* by Giorgio Vasari, which was first published in 1550, brought Renaissance ideas to the studios, drawing rooms and libraries of architects and patrons throughout Europe and the New World.

The ideology and theoretical integrity of Palladio's Renaissance was largely lost to a Europe that was still bound to medieval planning and the Gothic. The classical language of Renaissance architecture became a vocabulary of ornament and decoration, particularly in France under the patronage of François I (1494–1547), and the great chateaux that were built in the Loire valley were undoubtedly influenced by Leonardo da Vinci, who settled in France under the king's protection.

Spain's picturesque and highly decorated Gothic was fused with Italian ideas to create a style known as *Plateresque*, while in Portugal Manuel I gave his name to a similar movement known as *Manueline*. Although Henry VIII invited the Florentine artist Pietro Torrigiano (1472–1528) to Italianize his father's tomb in Westminster Abbey, London, the influence of Roman architecture on English buildings was restrained by

Protestantism. In France, on the other hand, the great chateau of Fontainebleau provided a home where the painters Giovanni Rosso (1494–1540) and Francesco Primaticcio (1504–1570) could develop a decorative style of high-relief stucco and painting, which developed under the direction of the Venetian architect Sebastiano Serlio (1475–1554) and the French master Philibert de l'Orme (*c*.1510–1570) into a French Mannerist/ Renaissance architecture that was more than a hybrid style.

Through trade, travel, commerce and religion, Franco-Italiante Mannerism spread

SANT'ANDREA

MANTUA, ITALY **LEON BATTISTA ALBERTI**

The church of Sant'Andrea in Mantua, Italy, was begun by Alberti in 1472, but it was not completed until the 18th century. Its plan – an aisleless nave and side chapels – became the pattern for many later churches. Alberti's façade echoes the triumphal arches of the Imperial Caesars, and the coffered barrel vaulting in the nave is a second homage to the architecture of ancient Rome.

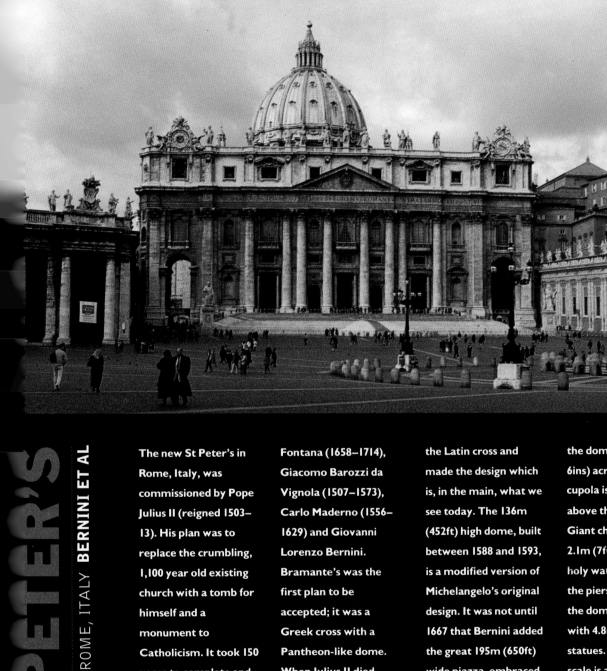

The new St Peter's in Rome, Italy, was commissioned by Pope Julius II (reigned 1503–13). His plan was to replace the crumbling, 1,100 year old existing church with a tomb for himself and a monument to Catholicism. It took 150 years to complete and employed the talents of the master artists and architects of the Renaissance – Donate Bramante, Giuliano (1443–1516), Antonio da Sangallo, Raphael, Baldassare Peruzzi (1481–1536), Michelangelo, Carlo

Fontana (1658–1714), Giacomo Barozzi da Vignola (1507–1573), Carlo Maderno (1556–1629) and Giovanni Lorenzo Bernini. Bramante's was the first plan to be accepted; it was a Greek cross with a Pantheon-like dome. When Julius II died, Raphael took over from Bramante, and proposed a Latin cross. When Raphael died, Peruzzi reinstated the Greek cross. Antonio da Sangallo tinkered with the plan after Peruzzi's death, then Michelangelo reinstated

the Latin cross and made the design which is, in the main, what we see today. The 136m (452ft) high dome, built between 1588 and 1593, is a modified version of Michelangelo's original design. It was not until 1667 that Bernini added the great 195m (650ft) wide piazza, embraced by the two, curved, colonnaded arms that ccntain 284 pillars. The most breathtaking aspect of St Peter's is its immense size; the nave is 25.2m (84ft) wide and 210m (700ft) long (including the portico)

the dome is 41.3m (137ft 6ins) across and its cupola is 100m (335ft) above the ground. Giant cherubs almost 2.1m (7ft) tall support holy water stoups, and the piers that support the dome are decorated with 4.8m (16ft) high statues. The awesome scale is a constant reminder that this is a building for the Almighty.

LITTLE MORETON

CHESHIRE, UK

WHO BUILT THAT?

50

Little Moreton Hall, Cheshire, UK, was built between 1550 and 1559. The distinctive black and white ("magpie") patterning of the half-timbering is typical of the English Midlands in Tudor times. The long gallery, 23m (75ft) in length, which projects over the walls of the main hall, was added in 1580. It is a nod to the Renaissance, which was slow in coming to England.

to all parts of Europe, particularly Hungary, Austria, Germany, Flanders (Belgium) and Holland. The close ties between the court of Elizabeth I of England and the Low Countries encouraged craftsmen from Holland and Flanders, carrying architectural pattern books by Hans Vredeman de Vries (b.1526) and Wendel Dietterlin (1551–99), to travel to England, where they influenced Elizabethan and Jacobean building. After the defeat of the Armada in 1588 the post-Reformation building saw a new simplicity in domestic architecture that was recognizably English and, ironically, it was distinguished by the prevailing absence of architects in favour of master-masons and carpenters. The great halls of Little

Moreton, Cheshire, Hardwick, Derbyshire, Blickling, Norfolk and Wollaton, Nottingham, were the apotheosis of a vernacular craft that was handed down from the pre-Reformation church builders of medieval England, to such 16th-century builders as Robert Smythson (1536–1614) whose skills were eclipsed only in 1573 by the birth of England's first architect, Inigo Jones (1573–1652).

The Queen's House at Greenwich, London, was the first Italian villa to be built in England. It was built by Inigo Jones, the great follower and popularizer of Palladio's work. Jones studied Palladian architecture at first-hand in Rome and Venice. Like the Villa Capra, the Queen's House is completely symmetrical, with pleasingly balanced proportions. It was begun in 1616 and was to have a profound influence on the neoclassical style in England.

QUEEN'S HOUSE

GREENWICH, UK **INIGO JONES**

ST PAUL'S

LONDON, UK SIR CHRISTOPHER WREN

St Paul's Cathedral, London, was built between 1675 and 1710 by Sir Christopher Wren (1632–1723). It is considered his masterpiece, even though it is technically an unsatisfactory essay in the English Baroque style. Wren was commissioned to build the cathedral (and 51 other churches) after the Great Fire of 1666 devastated the city of London. Arguments and differences of opinion dogged the project from planning stage to building. Wren originally designed St Paul's on a Greek cross plan and intended to build it in the new Baroque style he had seen in Paris the year before the fire. However, the clergy who commissioned him would have none of this modern nonsense and insisted on a traditional Latin cross with a long nave, aisles and transept. Wren had to adapt his plan, therefore, and the finished building is a compromise, with Wren making alterations on site. The dome, however, is pure Baroque; even so, the commissioners fought to the last. They insisted on the balustrade encircling the stone gallery at the base of the dome; Wren condemned it as frippery.

In spite of the troubles during its building, St Paul's has many outstanding features. Size is one of them: the cathedral was built to house a vast congregation, is 139m (463ft) long and covers an area of 5760m^2 (64,000sq ft). The vaulted crypt extends the whole length of the cathedral. The dome is an engineering feat: it is in fact three domes. The inner brick dome is a shallow, saucer shape with a central eye. Behind it is a conical dome, also made of brick, braced with iron chains and an iron band. This supports the lantern, the cross and the golden ball, as well as providing anchor points for the outer dome, which is made from lead covered timber.

Probably the most astounding thing about St Paul's is the man who built it: Sir Christopher Wren was not even an architect. He was a fellow of the Royal Society, an astronomer and a physicist – he regarded architecture as a hobby.

WESTMINSTER ABBEY

LONDON, UK **HENRY OF WESTMINSTER**

Westminster Abbey in London, defies architectural classification. It began as the abbey church of a Benedictine monastery founded in 960 by St Dunstan. It was partly rebuilt by Edward the Confessor just in time for William the Conqueror to hold his coronation in it (1066). From the Conquest on, English monarchs could not let it alone, and it has been refurbished, rebuilt, partly demolished, rearranged, added to and embellished so that it now displays architectural details from Norman, Gothic, Early English, Perpendicular, Tudor, Early Renaissance and Georgian periods. Much the most coherent contribution was made by Henry III and his master mason Henry of Westminster between 1245 and 1290. They took Rheims Cathedral as their model and achieved a vertiginous French loftiness and perpendicularity in the nave. Other architects involved in Westminster Abbey included Henry Yevele, John James, Nicholas Hawksmoor and Christopher Wren.

ROYAL HOSPITAL

GREENWICH, UK **SIR CHRISTOPHER WREN**

The Royal Hospital, now the Royal Naval College, at Greenwich, London, was built by Sir Christopher Wren between 1696 and 1713. Using the gracious Queen's House by Inigo Jones as a focal point and the King Charles block, which had been built by John Webb in 1663–7, as a starting point, Wren composed a symmetrical palace of great classical beauty. Queen Anne's block and blocks for King William and Queen Mary complete the composition. Colonnades and domes add restrained decoration.

The Baroque style in Germany and Austria was characterized by exuberant decoration. St Peter's Church, Vienna, Austria, was built by Johann Lukas von Hildebrandt (1688–1745), a German who was born in Italy, and trained there as an architect before becoming the engineer to the Austrian court in Vienna. His forte was ornamentation rather than originality, and the sumptuous interior of the church was much copied during the Baroque period that was inspired by the Counter-Reformation of the Catholic church after the challenge of Luther and the Protestant Reformation.

BAROQUE

While Europe celebrated the dawn of the 17th century with a new Baroque architecture that was to survive for 200 years, Jones followed his mentors, Alberti and Palladio, to Rome, where he studied neoclassical buildings in the company of his patron, the Earl of Arundel. His return to England led to an extraordinary paradox. While Europe had moved from the austerity of Bramante's classicism, through the French "Fontaine-bleau style", to the decorated sensuality of Baroque architecture, England emerged from a stone, timber and brickwork craft tradition to embrace an apparently revolutionary style, which, under Jones's hand, returned the Renaissance to the rigour and scholarship of the early period.

From 1618 until his death in 1652, Inigo Jones dominated architecture, and he left to the Stuart period of English history a new tradition of classicism or Palladianism, which challenged the Dutch-influenced brick and stone style and established a platform on which Sir Christopher Wren (1632–1723) continued to build. After a distinguished career at Oxford University as Professor of Astronomy, Wren was appointed surveyor-general of the King's Works. He was influenced by the French Baroque, a style that is evident in many of his great buildings, especially those designed after the devastations of the Great Fire of London in 1666. Wren's individuality, combined with the influences resulting from Charles II's earlier exile in the French court, gave England its Baroque period as an introduction to Georgian architecture.

The Palace of Versailles, the court of the Sun King, Louis XIV of France, began life as a modest hunting lodge. The shell built around it by Louis it by Louis le Vau and, later, Jules Hardouin Mansart transformed it into the quintessential French Baroque palace. The transformation, which began in 1669, took almost 100 years to finish. The French Baroque uses immensity of scale, a combination of painting, sculpture and architectural features, and richly decorated interiors, but the curved lines and broken forms of the Italian Baroque did not feature.

Another version of Palladio's Villa Capra, Chiswick House, London, was built by Lord Burlington and William Kent (1685/6–1748) in 1726. Although Lord Burlington, one of the leading proponents of neoclassicism, had studied Palladio's work and been to Italy, Chiswick House is not a carbon copy: it is not symmetrical, and only two sides are identical. Burlington also started the fashion for "improving" the landscape to make it one with the architecture of the house. The grounds at Chiswick were laid out to resemble Pliny's Tuscan garden.

CHISWICK HOUSE
LONDON, UK **LORD BURLINGTON AND WILLIAM KENT**

The term Baroque is similar to *Gothic* in that both were invented by 19th-century critics as pejorative words to describe a low or base architecture. Gothic – meaning of the Goths or barbaric – came to represent the style in which some of the greatest buildings on earth were constructed, and in the same way, the word Baroque – originally intended to mean irregular and misshapen and appertaining to the pearl industry – was transformed from a term of abuse into a collective noun for a Roman style that became a global symbol of 17th- and 18th-century architecture.

The decline of Mannerism coincided with the Counter-Reformation, powered by the ascetic Jesuit St Ignatius Loyala, which included the purging of heretics by the Inquisition and a Catholic revival that demanded a total reappraisal of the role of art and architecture. It was decreed that painting should be directly at the service of an austere church, which required the figurative and pictorial representation of miracles, martyrdom, apotheoses and ecstasies as a recognizable and descriptive record of Catholic doctrines. Architecture was reserved for the churches' glorification and transcendence, and architects, possibly for the first time, became concerned with symbol, metaphor and content, in addition to the academic question or architectonic aesthetics and style. The dramatic evangelism of the Jesuits required an architecture to suit – a

BRIGHTON PAVILION

SUSSEX, UK **JOHN NASH**

Brighton Pavilion, Sussex, UK, was once a modest country villa. Between 1815 and 1821, however, it was enlarged and enhanced by John Nash (1752–1835) to make a fitting love-nest for the Prince Regent, later George IV (1762–1830). Nash was the leader of the Picturesque School, which encouraged asymmetry, fantasy Gothic and exotica. Consequently, Brighton Pavilion bristles with pinnacles, cupolas, pagodas and onion domes without and shimmers with decorative chinoiserie within. Nash's whimsy is, however, firmly anchored: much of the pavilion's structure depends on cast-iron elements, which were first used here in a domestic setting.

In England the Baroque was seen at its best in secular buildings. One of the masterpieces of the style is Blenheim Palace, Oxfordshire, which was built between 1705 and 1724 as a reward for the Duke of Marlborough for his military prowess in France. At once monumental, palatial and fearfully symmetrical, Blenheim satisfies the Baroque demand for massive scale and heavy ornamentation. It was built by the gifted gentleman-amateur, Sir John Vanbrugh (1664–1726), in partnership with Nicholas Hawksmoor (1661–1736), who undertook the more onerous parts.

UNIVERSITY OF VIRGINIA LIBRARY

USA **THOMAS JEFFERSON**

Gothic host in Renaissance clothes – that could address drama and language simultaneously, and thus the fluent Mannerism of Giulio Romano gave way to the Baroque architecture of Gian Lorenzo Bernini (1598–1680), Francesco Borromini (1599–1667) and Pietro da Cortona (1596–1669). While Caravaggio (1573–1610) and the great Baroque painter Jacopo Tintoretto explored the religious potential in the art of chiaroscuro, architects indulged in equally dramatic lighting effects.

In 17th-century Rome the Baroque movement was distinguished by a homogeneity of the arts, similar to that which had prevailed in Ancient Greece. Sculptors introduced colour to their work with painting; architects used sculpture as a means of structural support; and painters decorated interiors with giant frescos of architectural perspectives.

There was a growing frustration with the constant reworking of the square, rectangle, circle and sphere. New forms were found, including the oval, which became popular, partly no doubt because of its geometrical elasticity, sensuality and spatial compatibility to the acoustic requirements of music played and sung by the great choirs of Monteverdi and Vivaldi. At St Peter's in Rome, Bernini, the greatest 17th-century sculptor, laid out a huge colonnade to surround the piazza that could be seen as the two arms of a powerful Christian church, embracing an area, which, in its heroism, was a metaphor for the exploration of the mysteries of space and infinity that were intriguing Galileo and Copernicus and astonishing the world with the view that the earth was not the centre of the universe, but merely one heavenly body among countless others.

A fine example of the classical revival in the United States, the library of the University of Virginia, in Charlottesville, was built between 1817 and 1826 by Thomas Jefferson (1743–1826), the third President of the USA. Like many buildings on the campus, it was built in the style of a famous classic example as a teaching aid for the students: the library was modelled on the Pantheon in Rome. The original building burned down at the beginning of the 20th century, but was rebuilt exactly by McKim, Read and White.

While Venetian architects remained loyal to the tenets of Palladianism, despite its strong Roman Baroque counter attack, France weakened. Louis le Vau (1616–70) built the Baroque chateau of Vaux-le-Vicomte and the Palace of Versailles in counterpoint to the orthodox Renaissance work of his great contemporary François Mansart (1598–1666), and le Vau's successor, Jules Hardouin Mansart (1646–1708) extended the Baroque vocabulary in his additions to Versailles.

The Catholic stronghold of Vienna was a natural arena for the drama of the Baroque, with the palaces and churches of Johann Bernhard, Fischer von Erlach and Lukas von Hildebrandt (1688–1745) being among the finest of the period to be built anywhere in the world. The wealthy princes and powerful monasteries of the Danube valley built palaces and churches wherever Baroque architecture spread to Poland, Hungary and Russia, and Italian architects carefully manipulating the almost limitless configurations of the architectural language to provide a syntax and grammar that could be skilfully tailored to the specific requirements of the Russian Orthodox church. At the drawing-table of Bernini's assistant, Francesco Borromini, it would seem that the application of the Baroque style had no limits. Borromini's spiral finial over the lantern of the star-shaped church for the University of Rome represented the quintessential moment of sculptural expressionism and architectural plasticity, beyond which it would seem that nothing could follow, except, perhaps, a return to the austerity of Vitruvian classicism or the structural and formal integrity of Gothicism. This is especially true in England, where the academic classicism of Inigo Jones found little favour until it was manifested in the work of Christopher Wren, James Gibbs (1682–1754), Thomas Archer (1668–1726) and Nicholas Hawksmoor (1661–1736). In an age of Descartes, Corneille, Poussin and Claude and in the spirit of an English scientific tradition championed by Wren, architecture was slow in changing. However, Wren's London churches, culminating in the masterpiece of St Paul's, all displayed an eclectic virtuosity of architectural composition and language, which combined the genius of his friend Bernini and the mathematical exactitude and rationalism of the contemporary philosophical opus and scientific excitement created by Sir Isaac Newton, co-founder, with Wren, of the Royal Society. Yet, perhaps appropriately, the mastery of English Baroque architecture fell to Sir John Vanbrugh (1664–1726), entirely untrained and therefore free from the academic traditions of aesthetic pedagogy and dogma. A writer of Restoration plays, Vanbrugh's love of the theatre is entirely suited to the working out of the Baroque, to which Blenheim Palace, Oxfordshire, and Castle Howard, Yorkshire, are his finest testimonials.

PARIS OPERA HOUSE

FRANCE **CHARLES GARNIER**

The Opera in Paris, built by Charles Garnier (1825–98) between 1861 and 1874, is a masterpiece of French neo-Baroque, and the building came to characterize the "Second Empire Style". Opulent, ornate and sumptuous – it was described by one disaffected contemporary as looking like "an overloaded sideboard" – it was a set piece in the new Paris of wide boulevards and elegant squares that Napoleon III (1808–73) and Georges Haussmann (1809–91) created between 1850 and 1870. Just as the wide boulevards, however

decorative and pleasant, had a more rigorous subtext (they were to enable the army to move quickly around the city to disperse insurgents), so Garnier's exuberantly ornamented masterpiece is underpinned by solid, rigorous symmetry. Napoleon III set a competition for a new Opera House after a failed attempt on his life outside the old one. Garnier had to contend with a difficult, cramped site, to deal with the logistics of an affluent, carriage-borne audience, to ensure security for the Emperor, and to make sure the building

fulfilled its primary function. Entrance steps, foyers and carriage ramps controlled the ebb and flow of the crowd; a separate entrance for the Emperor and his retinue on one side of the building was matched by second on the other side for patrons with season tickets. In the working area of the Opera House, ornament was not allowed to obscure function – a huge stage area could handle most works and a high tower accommodated all the scenery changes needed. Garnier's true genius was to understand that the real reason for going to

the Opera was to be seen. He provided his patrons with a gorgeous foyer, with painted ceilings, chandeliers and gilded statuary that would not have disgraced Versailles (the Opera was nicknamed the "Palais Garnier"), and a magnificent wide marble staircase, the *escalier d'honneur* where costume and person could be displayed to advantage. At once frivolous and grandiose, the Opera is now recognized as a masterpiece. The exaggerated style of its outer form perfectly indicates its function: opera is, after all grand stylish, flamboyant art.

BIBLIOTHEQUE STE GENEVIEVE

The library of Ste Génévière, Paris, was the first important public building to use cast iron and wrought iron as a major structural element. It was built between 1843 and 1850 by Henri Labrouste. Iron was used for the structural framework of the vaulted roof and its supporting pillars. However, the exterior was clad in a classical masonry shell. While looking to the future in terms of its structural design, Ste Génévière has a romanesque feeling, created by the arcades of rounded arches and its ecclesiastical groundplan of long "nave" with flanking "aisles".

ROCOCO

After the great days of Baroque, the High Renaissance, led by Bernini and Borromini, and followed variously by Mansart and le Vau in France, Fischer von Erlach and von Hildebrandt in Austria, Zimmerman in Germany, Churriguera in Spain, and Wren, Hawksmoor and Vanbrugh in England, and before a period of Revivalism, France emerged from the reign of Henri IV (reigned 1589–1610) to establish a wealthy bourgeoisie under the political patronage of high taste in the salons of country chateau and *hôtels*. In the next century, during the transitional period from Louis XIV (1638–1715) to the regency of his great grandson, Louis XV (1710–74), a demand for comfort, intimacy and ornament led to the late Baroque variant of Rococo.

The word Rococo derives from the French word *rocaille*, meaning sea rocks and shells, and it is applied to the highly ornamental and decorative strain of late Baroque architecture. Principally an interior refinement, the Rococo style was made possible by advances in plastering techniques, some of which travelled across the world in the 1703 edition of Joseph Moxon's *Mechanical Exercises*, which describes techniques and tools for plastering that were popularized by Robert Adam (1728–92) and his younger brother, James (1732–94). Their work was in a restrained and sophisticated neoclassical style, which signalled the 18th-century age of Revivalism as a predictable reaction against the golden age of German Rococo, a style that was misinterpreted as decadent, ostentatious and an appropriate icon for the collapse of the Baroque, itself described by the influential English critic John Ruskin (1819–1900) as "grotesque Renaissance".

HOUSES OF PARLIAMENT

However, the contribution made by the Rococo style to world architecture is important and twofold. First, by its apparent licentiousness, Baroque-Rococo raised the level of critical inquiry beyond Vitruvian, Albertian and Palladian theory, which held that the aesthetic value of architecture could be judged only by its truthfulness to the prescriptions of antiquity. By its very nature, it demanded that architects be masters of an holistic art that embraced sculpture, painting, craft and extraordinarily sophisticated constructional techniques, an achievement epitomized by Bavarian brothers Cosmas Damian Asam (1686–1739) and Egid Quirin Asam (1692–1750). Second, the Baroque-Rococo represented a stand for originality. This was evident in the work of the French eccentric Claude Nicholas Ledoux (1736–1806), whose dangerous relationship with the French aristocracy during the French Revolution and subsequent imprisonment caused him to write the treatise *L'architecture* (1804), which added fuel to the critical reaction against the Baroque in its claim that globes, cylinders and pyramids were the matrix of constructive art.

But the influence of Ledoux and his contemporaries in England and the rest of Europe was nothing in comparison to the global political events during the late 18th and early 19th centuries. The American War

The Gothic Revival of the 19th century was particularly vibrant in England. In London, the Houses of Parliament helped to establish neo-Gothic as the prevailing style for public buildings. Built between 1840 and 1865, its plan was devised by Sir Charles Barry (1795–1860), and its Gothic detailing and interiors were the work of Augustus Welby Northmore Pugin (1812–52), the popularizer of the style.

AMSTERDAM BOURSE

THE NETHERLANDS HENDRICK BERLAGE

The Stock Exchange, or Bourse, in Amsterdam was built between 1898 and 1903 by Hendrick Berlage (1856–1934). A simple, dignified building, it honours the robust honesty of its materials (brick, steel and glass) by leaving them exposed and undecorated. This early manifestation of high tech functionalism has led critics to describe the building as "proto-modern". It is saved from aridity by Berlage's insistence on the use of Romanesque masonry techniques, which means that the structure of the building also provides its ornament.

of Independence, supported by the French and Spanish, put an end to British colonial rule, ruined the French economy and exhausted the Spanish, after their own war with Britain in 1727–8. George Washington became the first President of the United States in 1789, the year that British convict settlements were established in Australia and Louis XVI bowed the establishment of a constitutional monarchy, which led, in turn,

to the Revolution and to the French Republic of 1792. While Catherine II reformed Russian law, Napoleon defeated Austria, to be defeated himself by Nelson in the Battle of the Nile, which indirectly resulted in the extension of British rule throughout India. In 1801 Alexander I became the Tsar of Russia, and in 1804–5 Napoleon assumed the titles of Emperor and King of Italy.

The Holy Roman Empire ceased to exist in 1806 when Francis II renounced the title Emperor of Austria, its fall being matched by the rise of the French Empire, which had ambitions to expand into Russia but was ultimately defeated by the Russian winter of 1812. Two years later, the Duke of Wellington forced Napoleon into abdication and exile on Elba, after which the French monarchy was restored under Louis XVIII, together with monarchies in Austria, Prussia and the Netherlands, with Ferdinand I regaining the Italian throne in 1815, the same year in which Napoleon was finally defeated in the Battle of Waterloo. Greece was liberated from the Turkish yoke and became established as an independent kingdom, while Spain lost Mexico and Peru, and Brazil became independent from Portugal.

All of these extraordinary events occurred, paradoxically, within an age of reason and enlightenment, which produced a synthesis of the arts and science in real and theoretical terms. Georg Hegel's philosophical work *The Science of Logic* (1812–16), which proposed a new method of critical inquiry into the nature of reason, was matched by the musical genius of Ludwig van Beethoven (1770–1837), who bridged the classical and romantic periods of a European art that was celebrated by the poet Percy Bysshe Shelley (1792–1822) and enacted the life of George Gordon, Lord Byron (1788–1824). Pure science in France combined with applied science in Britain to make possible the industrial revolution, and much of the concomitant wealth went into the patronage of painting, sculpture and architecture.

The Red House at Bexleyheath, Kent, UK was built by Philip Webb in 1859 for William Morris, the leader of the Arts and Crafts Movement in England. True to Morris's idealistic and socialist principles, and in contrast with the ornate, Italianate villas still being built, the house is informal and rustic. It was built in the local vernacular style, using simple materials and the skills of master craftsmen. Morris himself built the kitchen units. The high, pitched roofs and pointed arches are a gentle reference to the Arts and Crafts Movement's affection for medieval Gothic.

NEOCLASSICISM

As a response to the need for order after the chaos of revolution, colonizations, restored monarchies and political change, architects looked back, rather than forwards, for a style that could provide stability. In a philosophical and scientific age of reason and objectivity, the Jesuit Abbé Langier published the *Essai sur l'Architecture* (1752). Its authority combined with Colin Campbell's *Vitruvius Britannicus*, Leoni's *Architecture of Palladio* and William Kent's *Designs of Inigo Jones* (1744), to establish a neoclassical revival.

Eventually, however, the spirit of the individual, the romance of medieval Gothic and the exoticism that came from travelling in India, China and beyond, led to a bizarre chapter in the evolution of architecture, in which the 19th century saw the rich treasury of history being ruthlessly plundered in the name of eclecticism and stylistic revivals. Architecture was at the edge of aesthetic anarchy, but it was rescued by four very different persuasions. The British critic and commentator John Ruskin led a movement against the pastiche and parodies of 19th-century eclecticism to re-establish a moral and ethical code for architecture. William Morris (1834–96) proclaimed the dependency of beauty on truth and especially the truthfulness of materials to perform their function optimally. However, Ruskin was happier to cite the medieval Gothic period as a paradigm to his view of architectonic morality than contemporary innovations, such as the Crystal Palace built by Sir Joseph Paxton (1801–65) for the Great Exhibition in 1851.

is made of solid stone, but its unnerving organic shapes and sculptures look as if they have been hand moulded in soft clay and as if they might melt away in the next rainstorm. Built by the Catalan architect Antoni Gaudi (1842–1926), who began it in 1884, it is still unfinished, and there is no indication that it will ever be complete.

La Sagrada Familia in Barcelona, Spain, is an extraordinary and fantastic building that defies classification, although it is usually grouped with art nouveau architecture. It

ARTS AND CRAFT MOVEMENT

Morris gave Ruskin's position a materiality that eventually led to the highly influential Arts and Crafts Movement, which manifested itself largely by the visible appearance of things rather than their inner structure. This was left for another admirer of the Gothic, Eugène Viollet-le-Duc (1814–79). His analysis of medieval architecture was translated into a contemporary theory that approximates to a modern idea of functionalism, whereby beauty and the judgement of good architecture should depend more on the fitness of the structure and its construction to perform their functions and less on stylistic, decorative and ornamental systems.

The Viceroy's House in New Delhi, India, was built by Sir Edwin Lutyens (1869–1944) in 1912–31. In the early 20th century, the classical style was often favoured for official or public buildings, particularly those in the colonies where it was felt necessary to build impressive buildings, redolent of dignity and authority. Lutyens was a keen neoclassicist, but here modified the style to suit the context, with Indian-style decoration replacing Renaissance ornamentation.

the working classes and the subsequent demands for higher wages. Steel also liberated the art of the engineers, who were able to demonstrate the beauty of its lightness and strength in opposition to the ornamented ironwork preferred by art nouveau architects such as Victor Hora (1861–1947) and Hector Guimard (1867–1942).

Thus the void left by the lack of orginality inherent in the neoclassical, classical, romantic and Gothic revivalist schools of the mid-19th century was filled by a period of experimentation, the spirit of which created buildings as progenitors of the 20th-century Modern Movement.

The Arts and Crafts Movement eventually turned its back on the academic tradition and

A new fascination for machines came about as a product of the industrial revolution. The invention in 1856 by Sir Henry Bessemer of a steel producing process made possible the almost immediate mechanization of industry. in part a response to the new unionization of

concentrated instead on a re-interpretation of the English rural vernacular. Philip Webb's (1831–1915) design for the Red House at Bexleyheath for William Morris in 1858, Norman Shaw's ideas for a "garden suburb" at Bedford Park in London and C.F.A. Voysey's engagingly new arrangements of informal geometries and composition created a British domestic architecture that is still copied today. The German government went to the lengths of appointing a special attaché, Hermann Muthesius (1861–1927), to its embassy in London with the specific task of reporting on the new developments in English architecture, which culminated in the 1904 publication of *Das Englische Haus*.

The Eiffel Tower in Paris, France, was built for the Paris Exhibition of 1889 as a showcase for the engineering skills of the bridge-builder Gustave Eiffel (1832–1923) and as a celebration of the structural capabilities of steel and wrought iron, the materials that dominated the architecture of the late 19th century. It is 300m (984ft) high and caused as much outrage and controversy in its day as did Richard Rogers's (b. 1933) Pompidou Centre almost 100 years later. The Eiffel Tower is often classified as an art nouveau construction, but the decorative arches that link its four massive "feet" serve no structural purpose.

EIFFEL TOWER

PARIS, FRANCE **GUSTAVE EIFFEL**

GLASGOW SCHOOL OF ART

SCOTLAND **CHARLES RENNIE MACKINTOSH**

The Glasgow School of Art, built between 1896 and 1909, is often celebrated as the building that pioneered the Modern style in Europe. Its architect, Charles Rennie Mackintosh (1868–1928), is similarly celebrated as the first Modern architect in Europe. The School was built in two phases, after a competition to design it was set by the Board of Governors,

who offered a frugal £14,000 budget for the first phase and made it clear that they wanted a "plain building". The competition was won by Honeyman and Keppie, where the 28-year old Mackintosh was a junior architect. Although it was his design, his name did not appear on the drawings. However, by the time the go-ahead was given for the second phase in 1906, Mackintosh had

become the best-known and most avant garde of European architects. He had made several important new buildings including Hill House, Helensbrugh, Scotland and the Willow Tea Rooms, Glasgow and had exhibited work with his wife, Margaret Macdonald, at the Vienna Secession in 1900 and at the International Exhibition of Decorative Arts at

Turin in 1902. Consequently, he was able to drastically redesign the library and the West Wing of the art school, producing a design decades ahead of its time.

Mackintosh was a great influence on his contemporaries, including Antoni Gaudi, Frank Lloyd Wright, Hector Guimard and the Vienna Secessionists – all adherents of the art nouveau phenomenon. Art nouveau was an attempt to break free from the historicist treadmill of revivalism. It manifested itself in different ways under the hands of different architects. Mackintosh sought to redefine architecture in terms of function, internal reality and building techniques. He also believed that architectural coherence and integrity was impossible unless the architect took full responsibility for the design of decor, fixtures and fittings, much as Sir Norman Foster does today. Mackintosh himself designed everything for all his projects, from the foundations to the teaspoons.

ART NOUVEAU

The effect of Muthesius's work was to popularize an otherwise parochial and somewhat historicist picturesque style as a peer to the "new art" or art nouveau architecture that was appearing in France, Belgium, Austria, Germany and Scotland and that was itself removed deliberately from the classical academies in an attempt to find an art that symbolized the *Zeitgeist*. While the Arts and Crafts Movement represented a group movement that was bound to a pseudo-moralistic position promulgated by Ruskin and practised by Morris, the architects of art nouveau seemed to have no common purpose

The Secession House or Gallery was built in 1897–8 to house the work of artists who had broken away – seceded – from the traditional Viennese art establishment. The architect of the building, Josef Olbrich, was a major figure in the Secessionist Movement, Austria's response to the art nouveau tide that was sweeping Europe at the time. His small building is crowned with a dome of open metal work (known as the "golden cabbage") and with its teasing references to classicism could almost be described as "proto-post-modernist".

DER·ZEIT·IHRE·KVNST·
DER·KVNST·IHRE·FREIHEIT·

VER·ŞACRVM·

except to play out their individual visions of zoomorphic and biomorphic forms, the concentration and sophistication of which was to mature in the drawing-tables of Charles Rennie Mackintosh (1868–1928) in Glasgow and Joseph Maria Olbrich (1867–1908) in Vienna.

When it was built in 1923–4, Gerrit Rietveld's (1884–1964) building for Mrs Schröder in Utrecht, Netherlands, was the most modern house in Europe and a perfect example of *De Stijl* in action. *De Stijl* (*The Style*) was a group of Dutch architects and artists, which flourished between 1917 and 1931. They believed that pure, abstract geometry was the way to liberate design from the constraints of temporality. The design by Rietveld for the Schröder house, with its flat planes, geometric shape, primary colours and rational plan, is a perfect realization of their theories.

THE MODERN MOVEMENT

American architecture was influenced by the early English colonists and a multitude of immigrant nationalities, all of whose styles were homogenized by the building material available – timber. In the 18th century American houses were influenced by the classicism of Europe, but with timber imitating stone and clapboard replacing the imported East Anglian framing tradition to create such bizarre sub-styles as the southern Plantation Palladian. Slowly, North America developed its own architectural vocabulary under the authority of Henry Hobson Richardson (1838–86), whose education at the École des Beaux-Art in Paris caused late French Romanesque to find a home in Boston, Harvard and Chicago. Charles Sumner Greene (1868–1957) and his brother, Henry Mather Greene (1870–1954), developed a language derived from the Bengalese bungalow to bring the Arts and Crafts Movement to California, while Louis Henry Sullivan (1856–1924) and his protégé Frank Lloyd Wright (1867–1956) brought Chicago to the world's attention with a series of buildings that secured the position of Americans as key players in the evolution of the Modern Movement.

The Chicago school, the Arts and Crafts Movement, art nouveau, Dadaism, Futurism and an emerging radicalism among painters

ILLINOIS, USA **LUDWIG MIES VAN DER ROHE**

FARNSWORTH HOUSE

The Farnsworth House, near Plano, Illinois, USA was built in 1950 by Ludwig Mies van der Rohe. As an exercise in elegant minimalism it is hard to beat. There are no brick walls, pitched roof or solid foundations. A simple, single-storey, flat-roofed rectangular framework of white painted steel with plate glass walls, it floats on a concrete slab raised from the ground on low supports. A central core contains services (heating, plumbing) while the rest of the living space can be adapted to the owner's whim by using moveable partitions and fittings. The overall feeling is extremely Japanese, reminiscent of the buildings in the Katsura Palace.

WHO BUILT THAT?

VILLA SAVOYE

POISSY, FRANCE **LE CORBUSIER**

The Villa Savoye, Poissy, France, which was built in 1929–31, sums up Le Corbusier's architectural philosophy: it has a flat roof (a pitched one would spoil the geometry), which doubles as a terrace or garden; pilotis, or pillars, to lift the building free of the ground and to liberate the space beneath; an open-plan interior, which could be organized as needed, for all the load-bearing is carried on the structural frame; and ribbon windows, which are as continuous as possible. Everything works to provide light and space. Although the Villa Savoye has impeccable machine-age credentials – it could not have been built without reinforced concrete and mass-produced components – it has a graceful, sculptural quality, which later manifested itself in full bloom in Le Corbusier's masterpiece, the Pilgrim Chapel at Ronchamp.

PALAZETTO DELLO SPORT

ROME, ITALY **PIER LUIGI NERVI**

The Palazetto dello Sport, which has a capacity to seat 5,000 people, was built in 1959 for the 1960 Rome Olympics by the engineer Pier Luigi Nervi (1891–1979). Reinforced concrete is the signature material of the early and mid-20th century, when it took over from iron and steel as the preferred material for spanning large spaces. Nervi's work displays both its structural and its aesthetic qualities. The Palazetto dome is made of pre-cast, reinforced concrete coffers. The thrust of the roof, which is 12cm (4¾in) thick, is transmitted through 36 Y-shaped, concrete struts, the 20th-century version of flying buttresses.

and the literati all contributed to a late 19th-century atmosphere of experiment, revolution, polemic and critical inquiry. This intellectual activity coincided with the appearance of new materials such as reinforced concrete, glass and steel, and it was influenced by virtuoso works of engineering such as the Eiffel Tower (1889) and the Hall of Machines (1889). It is not surprising that architects such as Le Corbusier (1887–1965), Ludwig Mies van de Rohe (1886–1969) and Bruno Taut (1880–1938) among many others, were brought together under the educative

BAUHAUS
DESSAU, GERMANY WALTER GROPIUS

The Bauhaus at Dessau, Germany, was the college of art, architectue and design that was the cradle of Modernism. It was built in 1925 by Walter Gropius (1883–1969), who was also its principal. The severely functional and undecorated building contains workshops, domestic accommodation, lecture rooms, a refectory, a theatre, a gym, student accommodation and Gropius's own architectural practice office. The building was an uncompromising icon of the machine age – it was built entirely from prefabricated, reinforced concrete, mass-produced metal components and glass. Gropius encouraged students and staff from all over Europe, and many of the 20th-century's most important and influential architects taught there, including, apart from Gropius himself, Marcel Breuer, (1902–81), Laszlo Moholy-Nagy (1895–1946) and Ludwig Mies van der Rohe. In 1932 the Nazi government, whose racist fanaticism was unable to deal with internationalism, closed the school down. Mies van der Rohe and Gropius went to the United States, where they established Modernism as the style of mid-20th-century architecture.

LA NOTRE DAME DU HAUT

RONCHAMP, FRANCE **LE CORBUSIER**

The Pilgrim Chapel of Notre Dame du Haut at Ronchamp, France, is Le Corbusier's masterpiece. Built between 1950 and 1955, it is unlike any of his other buildings and appears to fulfil none of his celebrated "Five Points towards a New Architecture", set out in *Vers une Architecture* (1923): free, open plan; façade freed from the structure of the building; ribbon windows; roof terrace; pilotis to raise the building from the ground. Notre Dame has massive battered walls, so that there is no separation of the façade from the structure of the building; the ribbon windows have dwindled to a thin strip of daylight, made possible by raising the roof slightly on metal supports; the timber roof is not flat and will never accommodate a roof terrace; and the whole building crouches snugly against a hillside rather than perching on pilotis. However, some typical Le Corbusier characteristics have been retained: open planning has been achieved by putting the altar and pulpit *outside* the church, to accommodate 12,000 pilgrims; and the building material is rough cast concrete. With its massive, slabby walls, billowing roof and irregular, wedge-shaped windows filled with bright-coloured glass, Notre Dame du Haut is more a piece of sculpture than a building, a reminder that Le Corbusier was originally a painter. The chapel at Ronchamp, built towards the end of the architect's life, has been vilified by some critics who felt that Le Corbusier had betrayed the principles of the International Style he had helped to forge. Others, now the majority, see it as his most personal statement, a fitting epilogue to a career composed of controversy and creativity.

BRASILIA CONGRESS BUILDING

BRAZIL **OSCAR NIEMEYER**

The National Congress Buildings in Brasilia, Brazil, which were finished in 1960, were built by Oscar Niemeyer (b.1907). He was an avid disciple of Le Corbusier, whose building style – cool concrete slabs, *brises-soleil* (grilles to deflect fierce sunlight), buildings raised on pilotis to give shade below – worked particulaly well in hot climates. Brasilia, the new capital of Brazil, was created from scratch in the desert and thus gave unprecedented freedom to the architect, who responded by making a building that is an exercise in pure geometry. The twin towers house the Secretariat, the saucer-shaped construction is the Senate Chamber, and the shallow dome covers the Hall of Deputies.

SEAGRAM BUILDING

NEW YORK, USA **PHILIP JOHNSON AND LUDWIG MIES VAN DER ROHE**

The Seagram Building, New York, USA, by Philip Johnson (b.1906) and Mies van der Rohe (1886–1969) set the pattern for the Modernist tall, glass boxes that dominated the skylines of Chicago and New York from the 1950s. Following the prototype Lever Building (Skidmore Owings and Merrill, 1952), the Seagram Building features metal and glass curtain walling and severely regular windows. These skyscrapers were set back from the street and raised on pilotis to provide circulation space underneath them. The Seagram Building is 39 storeys high, a relative dwarf by today's standards.

GUGGENHEIM MUSEUM

NEW YORK, USA **FRANK LLOYD WRIGHT**

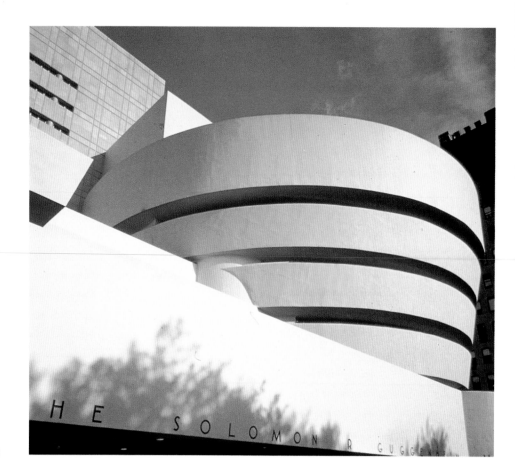

Built between 1943 and 1959, the Guggenheim Museum in New York, USA, was one of the last major works by Frank Lloyd Wright and a summation of the themes he had been working on all his life. Completely free from external ornament, its simple, organic shape follows a spiral ramp that gets wider as it rises. The outside of the building is a mirror of the inside. The use of smooth concrete is typical of Wright's work.

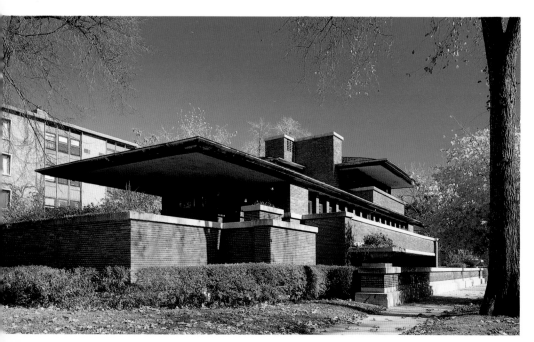

ROBIE HOUSE

ILLINOIS, USA **FRANK LLOYD WRIGHT**

Built in 1908–9 to fit a corner site in Woodlawn Road, Chicago, USA, the Robie House is one of Frank Lloyd Wright's earliest essays in what was to become his signature organic style. The strong horizontal elements and low, pitched roof keep what is, in fact, a two-storey building very close to the ground, almost as if it is growing up from it. The use of small bricks and concrete is typical of Wright's work.

FALLING WATER

PENNSYLVANIA, USA **FRANK LLOYD WRIGHT**

Falling Water, Bear Run, Pennsylvania, USA, which was built between 1937 and 1939, is Frank Lloyd Wright's most famous piece of domestic architecture. The horizontal span of its dramatic cantilevered balconies reflect the stone river bed below the waterfall. The use of smooth concrete and rough stone enhances the natural effect, making the house appear to have been carved from the living rock of the landscape rather than built in it.

CHANDIGARH HIGH COURT

INDIA **LE CORBUSIER**

The parliamentary buildings for Chandigarh, the capital of the Punjab in India were built between 1950 and 1965. They comprise the Capitol Palace, the Secretariat, Assembly and Supreme Court. Le Corbusier's largest and most ambitious project, it is at once majestic and functional, and bears the architect's unmistakeable mark. The main building material is exposed concrete, which is easily made on site by local workers. Long canopy roofs and *brises soleil* (sun shades) are integral to the buildings, which need protection from monsoon rains and fierce sun. Le Corbusier's style particularly suited countries with hot climates and his work greatly influenced Indian architects such as Charles Correa (b. 1930).

umbrella of the Bauhaus, itself a theoretical extension of the Deutsche Werkbund. Although English critics dismissed the new Germanic style as "modernisms", and found a strange ally in Hitler, who closed the Bauhaus in 1933, enough buildings had been built before World War II to establish a movement that, characterized as it was by a globally secured socio-political-functionalist agenda endorsed by the Congrès Internationaux d'Architecture Moderne, became known as the International Style. The pioneering years of 1908–33 were reinforced after 1945, when a huge rebuilding programme created a social role for architecture and town planning.

DOUGLAS HOUSE

HARPER SPRINGS, MICHIGAN, USA RICHARD MEIER

Built by Richard Meier (b. 1934) between 1971 and 1973, the Douglas house is built firmly in the tradition of Le Corbusier. All the elements are there – independent structural frame, flexible interior space, flat terraces and roofs, wrap around windows, sparkling white walls. However, the position on a wooded slope overlooking the water is pure Frank Lloyd Wright.

The emphasis on building legislation to control new public work developments, the technical requirements of industrialized building systems and partnerships with property developers to provide enlarged and new towns for the post-war population explosion has meant, however, that much of the architecture of the mid-20th century was carried out by public agencies in response to a social need, rather than by individual virtuoso architects searching for an aesthetic idealism. Yet in Italy, France, Scandinavia and the USA in particular, buildings by Le Corbusier, Pier Luigi Nervi (1891-1971), Erik Gunnar Asplund (1885–1940), Alvar Aalto (1898–1976), Hans Scharoun (1893–1972),

Berthold Lubetkin (1901–1990), Mies van der Rohe and Frank Lloyd Wright have reminded us that Modernism, in both its rational and romantic forms, was simultaneously functional and beautiful but unfortunately prone to being indifferently mimicked, a fact that was ultimately to trivialize and lay open to ridicule the intentions of the Modern masters.

So, while the pilgrimage chapel of Notre Dame du Hant at Ronchamp, finished in 1955 by Le Corbusier, and the Guggenheim Museum, completed four years later by Frank Lloyd Wright in New York, were applauded as works of genius, many critics saw the mediocrity of lesser architects in the 1960s and 1970s as an opportunity to bring down the International Style with a salvo of rhetorical invective that was equal in power to the dynamite used to demolish a failed icon of Modernism at the housing estate of Pruitt-Igoe in St Louis.

Le Courbusier's (1887–1965) Cistercian Monastery of Ste Marie de la Tourette at Eveux-sur-l'Arbresle was built on a traditional monastic plan (U-shaped building round a central court), it is a plain block realised in exposed concrete without luxury or adornment, and very suitable for its function. A mature, vigorous, confident work, built between 1955–1959, La Tourette has typical Le Corbusier features such as the flat roof and cantilevered block, which houses the monks' cells. However, the building is not raised on *pilotis* but built into the hillside. The exposed concrete used in this building bears the casting marks, textures and patterns made by the shuttering that formed it; Le Corbusier called it *breton brut* (raw concrete), which gave its name to "Brutalism", the style of architecture which used concrete for everything, regardless of suitability.

LA TOURETTE
EVEUX, FRANCE **LE CORBUSIER**

POSTMODERNISM

For a while architecture was left without a theoretical mentor. The recession in Europe in the 1970s underlined the level of architectural poverty, until the American architect and scholar Robert Venturi (b. 1925) published the book *Complexity and Contradiction*. Venturi called for an inclusive rather than exclusive architecture, and, in particular, an architecture that by referring to, and including elements of, history could communicate to a public that had been ostracized from the elitist and abstract architectural language of Modernism. Venturi's clarion call was followed in 1977 by Charles Jenks's book *The Language of Post-Modern Architecture*, which called for a "populist-pluralist art of immediate communication". With the help of American architects Philip Johnson (b. 1906), Charles Moore and Michael Graves (b. 1934), Postmodernism developed as a style of historical pastiche, metaphor, figurative representation and parody. But, like art nouveau, which also failed after only a short time, Postmodernism lacked an intellectual tradition on which it could be based.

Subsequently, "high-tech" architecture with a structural history starting with the Gothic, an intellectual base within the scholastic dynasty of Labrouste, Gaudet, Choisy, Perret and Le Corbusier, and an aesthetic tradition that developed from the first machine age following the industrial revolution through the Arts and Crafts Movement to the Bauhaus, will continue to sustain the reputation of architecture as it struggles through another period of classical and vernacular revivalism. The late 20th-century period of stylistic uncertainty will require more than a royal blessing if the great art of architecture is to celebrate the end of the second millennium with the same courage, originality and enthusiasm that greeted the end of the first.

LEICESTER UNIVERSITY

LEICESTER, UK **JAMES STIRLING AND JAMES GOWAN**

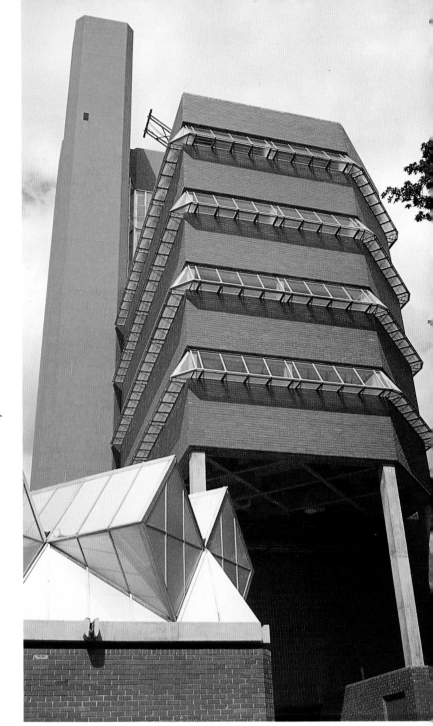

In England in the 1960s there was a sudden increase in building as the government began to invest in higher education and establish new universities. One of the most famous and individual responses was the Faculty of Engineering at Leicester University, built between 1959 and 1963 by James Stirling and James Gowan. Boldly poaching the high-rise language of the commercial sector, Stirling and Gowan used glass, brick, concrete and tiles to produce an uncompromisingly modern, forward-looking building, a symbol of the optimistic feeling that pervaded the decade.

Sydney Opera House, Australia, designed by Jørn Utzon (b. 1918) and built by Ove Arup (1895–1988) between 1959 and 1973, appears to float on the waters of the harbour, defying categorization. The materials used to build it (concrete, steel and glass), proclaim it to be a 20th-century building, but its form is unique. The great "sails" are made of concrete; their sheen comes from the white ceramic tiles that cover them. The spectacular design caused controversy at the time of building: the architect resigned it and the job was finished by Hall, Todd and Littlemore.

SYDNEY OPERA HOUSE

AUSTRALIA **JØRN UTZON AND OVE ARUP**

HONG KONG AND SHANGHAI BANK

full height of the building, the internal space is flexible to suit the changing demands of the client, the design is built around prefabricated elements and glass and steel are the preferred materials. The colours are muted and subtle, and structural elements such as the giant repeating "Xs," which hold the tie bars in place, are both decorative and functional. It also

features the sunscoop – an idea borrowed from the ancient Egyptians. The pyramids were lit by sunlight reflected from pools of water. The Foster sunscoop is a bank of mirrors outside the building which pick up sunlight, reflect it onto a second set of mirrors within, from where it is beamed down through the atrium and the glass floor of the building to the plaza below.
It took six years to build

the bank and there was much controversy along the way. To make room for the new building, the old bank, a monumental structure built in 1935, had to be demolished. As it had been a solid symbol of the colony's prosperity – the popular idea was that the bank prevented gold flowing down from the mountain into the harbour – the new building was not viewed favourably at first.
Foster faced a unique challenge in using western techniques in an eastern context. Elements that a western architect does not normally have to consider were very important: for example, the *fung-shui* had to be right, or bad luck would blight the bank. *Fung-shui* (windwater) is the Chinese method of interpreting the surrounding landscape in terms of its effects on the fortunes of those living and working in it. Foster employed a *fung-shui* expert, and the building is sited with the mountain behind it, to ward off winds and evil influences, and an unhindered view across the bay, to let in good fortune.

The headquarters of the Hong Kong and Shanghai Banking Corporation, Hong Kong, was built by Foster Associates (now Sir Norman Foster and Partners) who won the commission in 1979. Considered to be Foster's masterwork so far, it is an elegant, grown-up, high-tech classic. It is not a solid tower, but three slabs of different heights sandwiched together. Functional and skeletal elements are on view: the atrium reaches the

AT & T BUILDING

NEW YORK, USA **PHILIP JOHNSON**

The **AT & T Building** in New York, USA, by Philip Johnson and built between 1978 and 1984, ushered in the age of Postmodernism. Perhaps in reaction to the severe glass-curtained sheds of the Modernists (many of them built after the style of Johnson's own Seagram Building), the 1980s saw an outburst of frivolity, with architects taking classical elements and bright colours and applying them in a witty or jokey way to serious buildings. The AT & T Building, with its mammoth broken pediment on top of an undistinguished brick and glass skyscraper, plays tricks with scale and subverts the observer's expectations. Is it a real building or a giant, glass-fronted bookcase

VITRA CHAIR MUSEUM

WIEL AM RHEIN, GERMANY **FRANK GEHRY**

The complex of factory, offices and museum for the Vitra Chair Company was finished in 1989 by Frank Gehry (b. 1929). It is the ultimate in "post-post-modernism"; a deconstructed building in which all the elements are taken apart and left in a deliberately incoherent, chaotic heap of fractured shapes in a landscape. Colour is a unifying factor and the sculptural shape of the individual building elements is pleasingly organic.

HOUSE AT NEW CASTLE

DELAWARE, USA **ROBERT VENTURI**

The house at New Castle, Delaware, USA was built by Robert Venturi in 1978. Although Venturi himself eschews the label post-modernist, the house exhibits all the signs of the style: contradiction, dislocation, distortion of scale and an ambiguous, playful use of historical styles. The colonnade of fat, doric cut-out columns which fronts the house is sliced off at one side, presumably to make it clear that this is a style pun, not a piece of genuine kitsch.

SAINSBURY CENTRE

NORWICH, UK FOSTER ASSOCIATES

The Sainsbury Centre for the Visual Arts at the University of East Anglia, Norwich, UK, was built by Foster Associates in 1977. It is an early essay in what came to be known as "hi-tech" architecture. The essence of "hi-tech" is to display the structural elements of a building in an aesthetically pleasing manner. Glass and metal are the preferred media, and the grammar of industrial architecture is the preferred design framework. The Arts Centre was conceived as a giant shed in an open site – a kind of art hangar.

TWA BUILDING

NEW YORK, USA EERO SAARINEN

In the TWA building at Kennedy Airport, New York, concrete reaches sculptural heights only hinted at in Le Corbusier's later works. It was built by the Finnish architect Eero Saarinen (1910–1961) between 1956 and 1961. Expressionist rather than functional, the swooping, curving, undulating shapes show what concrete is capable of and make an easily-read essay in flight symbolism.

STAATSGALERIE

STUTTGART, GERMANY JAMES STIRLING AND MICHAEL WILFORD

The extension to the Staatsgalerie in Stuttgart, Germany, was built by the British architects James Stirling (b. 1926) and Michael Wilford between 1977 and 1984. Standing next to the existing gallery, which was built in the classical style, the new extension excited much controversy. The bright colours, exposed structural elements and brick and glass construction are Postmodern tropes, but the lack of ironic historicist references and the "style jokes" proclaim it to be a building of serious intent. The sinuous, organic flow of the building characterizes the later work of these architects.

WHO BUILT THAT?

91

AA: Architectural Association, London, leading progressive school of architecture and association with worldwide membership.

Abacus: the top slab of a CAPITAL.

Acanthus: the plant whose leaf is used as part of the decoration on column capitals of the Corinthian and Composite orders.

Acroterion (pl *Acroteria*): PLINTH at the apex of a PEDIMENT on which stands a statue or an ornament.

Adobe: Mud brick baked in the sun, commonly used in Africa, Spain and Latin America.

AIA: American Institute of Architects, Washington DC, USA.

Aisle: a walkway in a church or hall which runs laterally through the building usually parallel to the central nave.

Ambulatory: circulation or processional space or corridor surrounding the sanctuary of a religious building.

Apse: semicircular or polygonal end of the CHOIR in a church.

Arcade: a number of arches carried on piers or columns, or a covered passage lined with shops on one or both sides.

expressed universal values.

Ashlar: large, smooth-finished masonry blocks.

Atrium: originally a central room or court found in classical houses. In modern architecture the term is applied to a tall, multi-storey open central area within a building that receives light from above.

Axial Plan: a term used when a building is planned longitudinally, i.e. not a CENTRALIZED PLAN.

Axonometric: a geometric drawing around an axis which shows a building three dimensionally.

Baldachin/Baldacchino: a canopy over doorway, altar or throne.

Balloon Frame: a method of building using a timber frame commonly found in North America and Scandinavia. The construction begins with an enclosure of upright posts against which are nailed horizontal timbers.

Baluster: a short post or pillar supporting a rail which in a series becomes *balustrading*.

Baroque: exuberant, ornate architecture and interior design which proliferated in Europe *c.*1600–1750.

Basilica: in Roman architecture a rectangular building, usually a meeting hall, with central nave, two side aisles and often also including galleries. Today the term is applied to a church

Arches

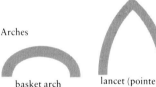

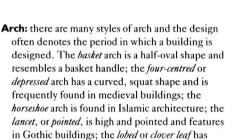

four-centered (depressed) arch · horseshoe arch · basket arch · lancet (pointed) arch · Tudor arch · Roman arch · Lobed (clover leaf) arch · ogee arch

Arch: there are many styles of arch and the design often denotes the period in which a building is designed. The *basket* arch is a half-oval shape and resembles a basket handle; the *four-centred* or *depressed* arch has a curved, squat shape and is frequently found in medieval buildings; the *horseshoe* arch is found in Islamic architecture; the *lancet*, or *pointed*, is high and pointed and features in Gothic buildings; the *lobed* or *clover leaf* has several concave segments; the *Roman* has a semi-circular top; the *ogee* consists of four arcs resembling an upturned boat keel and was introduced in around 1300; a *Tudor* arch is squat in shape and rises to an apex in the centre.

Architrave: the lower part of an ENTABLATURE running along the top of the row of columns. It is also commonly used to describe the moulding surrounding a door or window.

Arcuated: a building whose construction is based around the arch shape in contrast to the POST-AND-LINTEL type.

Art Deco: a style of architecture and decoration named after the 1925 Paris Exposition des Arts Décoratifs, epitomized by use of cubic shapes and angular, dynamic ornamentation.

Art Nouveau: general terms to describe flowing, sinuous designs based on natural forms, popular in Europe *c.*1895–1906. See JUGENDSTIL, SECESSIONSTIL.

Arts & Crafts: movement for design reform, initially centred around William Morris, which developed in England in the second half of the 19th century and was widely influential. In architecture, the movement aimed to abolish historicism and revivalism, and design "honest" buildings that

with a nave and two or more aisles.

Bastion: projecting part of a fortification as a look-out.

Bauhaus: progressive and radical German school of art and design, headed by architects Walter Gropius, Mies van der Rohe and Adolf Meyer, situated in Weimar, Dessau and Berlin, 1919–32. Exercised an important influence on the development of MODERNISM.

Bay: the vertical division of a building into segments usually denoted by columns or windows.

BDA: Bund Deutscher Architekten (German Association of Architects).

Beam: horizontal, load-bearing members of a building.

Beaux Arts: rich NEO-CLASSICAL style favoured by the Paris-based Ecole des Beaux Arts in 19th century France and well in to the 20th century.

Boss: ornamental projection at the intersection of ribs or beams.

Bracket: a small supporting piece of stone, wood or metal projecting from a wall.

Brise Soleil: a sun-break, usually or horizontal or vertical slats, used to shade a window.

Brutalism: architecture where the rough constructional materials are left exposed; term derived in part from use of *in-situ* concrete (*béton brut*) by Le Corbusier and others.

Buttress: masonry or brickwork which is built against, or projects from, a loadbearing wall to add strength (see FLYING BUTTRESS).

Byzantine: art and architecture of Byzantium (Constantinople); term used for eastern Christian art of the 4th–14th century.

Campanile: a bell tower, usually free standing.

Cantilever: a horizontal projection from a wall, often balcony or stair, supported at one end only.

Capital: top section or crowning feature of a column, often elaborate. The shape and design denotes the architectural order, e.g. DORIC, CORINTHIAN.

Carolingian: art based on a Roman revival, which appeared from the 8th to the 10th century in Charlemagne's empire – notably the Netherlands, France and Germany.

Caryatid: statue of a draped human figure, usually female, used as a supporting column.

Centralized Plan: building design radiating from a central point.

Chamfered: bevelled edge of stone or wood.

Chancel: part of church in which the altar is placed, also a term used to refer to the entire east end of a church beyond the CROSSING.

Chicago School: group of architects in Chicago which in the years 1880–1910 invented the steel-frame, multi-storey skyscraper.

Choir: portion of the church to the west of the altar used by singers and clergy.

CIAM: Congrès Internationaux d'Architecture Moderne – a liberal, reformist organization of architects, artists and designers founded at La Sarraz in 1928 and which supported MODERNISM. It disbanded in the 1950s.

CICA: Comité Internationale des Critiques d'Architecture, founded in Barcelona in 1979.

CIRPAC: Comité Internationale pour la Résolution des Problèmes des l'Architecture Contemporaine, a group within CIAM, founded in 1929.

Citadel: a fort usually placed at the corner of a fortified town.

Cladding: non-loadbearing external envelope or covering to a building.

Classical: the architecture originating in ancient Greece or Rome, the rules and forms of which were revived to establish the classical RENAISSANCE in Europe in the 15th and 16th century.

Clerestory: range of windows constituting the upper stage of a church above the AISLE.

Cloister: a space, usually a quadrangle, surrounded by a COLONNADE or ARCADE opening on to a central courtyard.

Coffering: recessed panels usually found in ceilings, VAULTS or domes.

Colonnade: line of columns bearing arches or a horizontal ENTABLATURE.

Colossal Order: see GIANT ORDER.

Column: vertical, round pillar or support for an ARCH or an ENTABLATURE.

Composite Order: see ORDERS.

Constructivism: radical arts and architecture movement in post-revolution Russia of the 1920s, concerned with abstract, geometric forms and utilitarian socially orientated design.

Corbel: a projecting BRACKET, usually of stone, to provide support for a beam.

Corinthian Order: see ORDERS.

Cornice: top section of an ENTABLATURE also used to describe the decorative moulding along the top of a building or wall.

Course: horizontal layer of bricks or stone forming part of a wall.

Crenellation: parapet or battlements with openings, usually at the top of castle walls.

Crossing: place in a church where nave and transept intersect.

Cross-vault: see VAULT.

Cruciform: cross-shaped.

Crypt: underground chamber of a church.

Cupola: a domed VAULT crowning a roof.

Curtain Wall: in modern architecture, the non-loadbearing skin or cladding on the outside of a building.

Deconstructivism: term derived for the contemporary philosophical/literary movement centred around Jacques Derrida. In architecture it is exemplified by exploded architectural shapes as in the "Deconstructivist" show in New York in 1987.

Decorated: term for late 13th/early 14th century British architecture characterized by ornate decoration and the ogee (S-curve) in arches and window TRACERY.

Dentil: small square block used at regular intervals to form part of a CORNICE.

De Stijl: Dutch political and creative anti-tradition movement founded by Van Doesberg in Leiden in 1917 (until 1931), whose work is characterized by geometric abstraction, the use of primary colours and the interplay of flat planes.

Domical Vault: see VAULT.

Doric Order: see ORDERS.

Drum: vertical wall, usually circular, on which sits a dome.

Early English: term for English GOTHIC of the 12th-late 13th century.

Eaves: the projection of a roof overhanging the walls.

Elevation: the vertical face of a building or a drawing showing a face.

Elizabethan: Late 16th–mid-17th century English architecture characterized by symmetrical façades, decorative STRAPWORK and large windows.

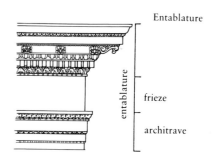

Entablature: section of upper wall or horizontal beam supported by columns consisting of ARCHITRAVE, FRIEZE and CORNICE.

ETH: Eidgenössische Technische Hochschule (Swiss Federal Institute of Technology), Zurich.

Expressionism: German movement in the arts which flourished between 1910 and 1924. Its architecture is typified by bold, sculptural, monumental buildings and bizarre Utopian projects.

Façade: the exterior face of a building.

Fascia: a plain horizontal band usually found in the ARCHITRAVE of an order. Also, a board or plate covering the end of roof rafters.

Fenestration: the arrangement of windows in a building.

Feng-Shui: the ancient art of living in harmony with nature and of planning mankind's relationship with the land. Much used by Chinese architects, builders and developers in the layout of their buildings.

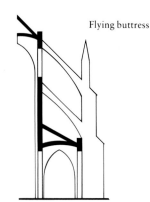
Flying buttress

Ferro-Concrete: steel-reinforced concrete.

Finial: formal ornament on top of a gable, pinnacle etc.

Flamboyant: 19th-century term for late French GOTHIC architecture of the 15th–16th century characterized by flame-like TRACERY and elaborate carving.

Fluting: vertical, concave channels cut into column shafts.

Flying Buttress: an arch or half-arch transmitting the thrust of a VAULT or roof from the upper part of a wall to a support outside.

Folly: a decorative building, often in the form of a classical ruin, used to enhance a landscape.

Fresco: method of painting directly on wall or ceiling before plaster dries.

Frieze: a horizontal band, often close to the top of a wall or forming the central part of an ENTABLATURE, usually enriched with sculpture or decorative paintwork.

Functionalism: basic MODERN MOVEMENT idea that the purpose, or function, of a building can be expressed through its utilitarian plan and unadorned external forms. Derived from German term *Zweck.*

Futurism: a European but Italian-based modern art movement originating in the "Speed and War" manifesto published by Marinetti in 1909. Architecture was added to the movement's arsenal in 1914 with Antonia Sant'Elia's "Futurist Architecture" manifesto.

Gable: triangular part of a wall at the end of a roof, also a triangular canopy above a door or window.

Garden City: idea formulated by Ebenezer Howard and others and first realized at Letchworth, Herts, UK, in 1903.

GATEPAC: Grup d'Artistes i Tècnics Catalans per el Progrés de l'Arquitectura Contemporànea. Spanish architects group founded by Sert and others associated with CIAM in 1930.

Georgian: the brick and STUCCO urban CLASSICAL-style language in England from the early 18th to the early 19th century.

Giant Order (also called Colossal Order): can apply to any ORDER whose columns climb to several storeys high.

Gothic: general term for medieval architecture characterized by pointed stone arches, rib VAULTS, great windows of coloured glass and skilful construction to create delicate-looking structures.

Gruppo 7: group of 7 Italian architects founded in 1926, later known as MIAR.

Half-timbered: building style consisting of an exposed timber frame with the gaps between filled with brick or plaster.

Hellenistic: architecture which developed between 323 BC and 27 BC in the kingdoms, including Greece and northern Egypt, which made the empire of Alexander the Great (356–323 BC). It is typified by rich decoration and sculpture.

High-Tech: a term employed for some architecture dating from late 1970s onwards which displays the technology of supports and services of a building.

Hipped Roof: a roof composed of inclining sides and ends.

IAA: International Academy of Architecture.

IBA: German abbreviation for the International Building Exhibition organization to renovate and rebuild West Berlin, 1977–87.

IIT: Illinois Institute of Technology.

International Style: An American term introduced by Hitchcock and Johnson to describe European MODERNIST buildings of the early inter-war years, 1920–31.

Ionic Order: see ORDERS.

Isometric: accurate, 45°-drawn three-dimensional projection.

Iwan: a term used of Near Eastern and Islamic architecture meaning a rectangular, vaulted hall enclosed by either two or three walls and often leading to a courtyard.

IUA: International Union of Architects.

Jugendstil: German term for ART NOUVEAU, which can be translated as "Youth Style". Its characteristics include naturalistic, whiplash ornamentation.

Keep: the main tower of a castle, often containing living quarters and storage space to be used in times of siege.

Keystone: the central stone of a semi-circular arch.

Lantern: small glazed turret on the top of a roof or dome.

Lattice Window: a window with diamond shaped panes of glass.

LCC: London County Council.

Lintel: wooden or stone beam across the top of a door or widow to support the weight of the wall above.

Madrasa: a Muslim school of theology attached to a mosque.

Mannerism: High RENAISSANCE tendency (beginning with late Michelangelo and popular throughout 16th century Italy, Spain and France) to break the Classical rules, transforming the Renaissance architectural language into a more individual style.

MARS: Modern Architectural Research, a British group of architects associated with CIAM founded in 1933 by Maxwell Fry and Wells Coates.

Metabolists: group formed in Tokyo in 1960 whose original members included Kenzo Tange and Kisho Kurokawa. Their projects are characterized by the use of futuristic, mechanical imagery, often using "pods" or cells, in order to "reflect dynamic reality".

Mihrab: a niche, used for prayer and facing Mecca, in a mosque wall.

MIT: Massachussetts Institute of Technology.

Modern Movement *or* Modernism: general terms to describe the new, socially progressive, undecorated, cubic, democratic and functionalist architectural intentions of the first half of the 20th century.

MOMA: Museum of Modern Art, New York.

Mullion: Vertical post used to divide a window.

Muqarna: see VAULT.

NATO: Narrative Architecture Today. Group within the AA founded by Nigel Coates.

Nave: the central, broad walkway in a church. The term is also applied to the entire western section of a church.

Neoclassicism: revival of the purity and formality of classical design (mid-18th–early 19th century) as a reaction against BAROQUE and ROCOCO extravagance.

Neue Sachlichkeit: roughly translated as the "new objectivity"; the post-Weimar Republic MODERNIST tradition associated with FUNCTIONALISM.

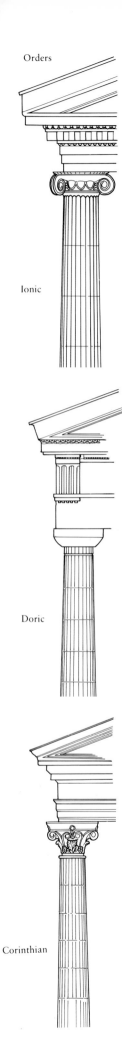

Orders

Ionic

Doric

Corinthian

Norman: form of ROMANESQUE architecture in England around the time of the Norman Conquest.

Obelisk: tall, tapering stone shaft finished with pyramidal top.

Octastyle: PORTICO with eight columns.

Ogive: see ARCH.

OMA: Office of Metropolitan Architecture of Rem Koolhaas.

Orders: term describing the classical architectural language of religious and secular buildings in ancient Greece and Rome. There are five principal orders – Tuscan, Doric, Ionic, Corinthian and Composite and each is denoted by a column with decorated ENTABLATURE. The *Tuscan* is primitive looking and derives from ancient Etruscan temples; the *Doric* is more refined than Tuscan and has its origins in ancient Greece but is also found in ancient Roman building; *Ionic* originated in Asia Minor in the mid-6th century BC and is distinguished by VOLUTES at the column tops and DENTILS in the CORNICE; *Corinthian* is thought to be an Athenian invention of around the 5th century BC and is distinguished by ornate ACANTHUS leaf decoration at its crown; the *Composite* is the latest and most elaborate of the orders and combines features from the Corinthian and Ionic orders.

Organic: an architecture closely associated with nature, in terms of landscape, setting, materials and often forms of construction. Used widely in relation to 20th century "individualist" and "democratic" thinking by Frank Lloyd Wright and his followers.

Palladianism: style popular in England in early 17th century, and then in USA, inspired by the work of Andrea Palladio (1508–80).

Pediment: the low-pitched triangular GABLE above an ENTABLATURE in classical architecture. In RENAISSANCE architecture and later it refers to any roof end whether triangular, semi-circular or broken.

Pendentive: concave structure, usually composed of inclining arches, which supports circular dome over square or polygonal base.

Peristyle: a line of columns surrounding a building or courtyard.

Perpendicular: late English GOTHIC architecture characterized by tall, stressed verticals, slender supports and large windows.

Pier: the solid, structure-carrying areas or pillars between windows, doors or other openings.

Pietra Dura: decorative work using inlaid semi-precious stones to depict geometric patterns etc.

Pilaster: shallow, ornamental, rectangular pillar projecting from a wall.

Pillar: a vertical support which differs from a column in that it does not have to be cylindrical nor conform to the orders.

Pilotis: term coined by Le Corbusier to describe pillars or stilts upon which a building is raised.

Plan: the layout of a building drawn in a horizontal plane.

Plinth: the square base upon which a column sits.

Podium: continuous base supporting columns.

Pointed Arch: see ARCH.

Porte-Cochère: an entrance large enough for wheeled vehicles to pass through.

Portico: covered entrance or centrepiece to a building, often supported on columns.

Post-and-Lintel: also known as post-and-beam, a construction method consisting of horizontal beams, or LINTELS, supported on vertical posts.

Postmodernism: mid-1970s reaction to MODERNISM seen as an historical and pictoral style; uses elements from different periods eclectically and sometimes light-heartedly.

Prairie School: architecture of the American Midwest, particularly Frank Lloyd Wright and contemporaries at the turn of the 20th century, which derived its inspiration from the open flat landscape and natural materials.

Prefabrication: manufactured components or entire buildings which are easily transported from factory to site. "Prefab" is the name given to factory-made emergency short life homes, particularly in the UK after the Second World War.

Pylon: in ancient Egyptian architecture, the large, chunky pyramidal towers flanking a temple entrance.

Quadratura: wall or ceiling painting which deceives the eye into believing objects portrayed are three-dimensional. In the 17th-18th century the work was undertaken by travelling painters known as "quadraturisti".

Queen Anne: late 19th century style borrowed from 17th century brick-built Dutch housing, typified by the designs of Norman Shaw.

Rampart: a stone or earth wall of defence around a castle or fortified town.

Reinforced Concrete: concrete (a mixture of sand, aggregate and cement) which is given additional tensile strength by incorporating steel rods within its mass.

Renaissance: rebirth of classical learning in architecture, beginning in mid-15th century Italy, which revolutionized building practice and took inspiration from classical buildings and writings.

Render: the application of STUCCO or cement mortar to the face of a wall to give a continuous smooth finish.

Rib: a projecting band on a ceiling or VAULT.

RIAS: Royal Incorporation of Architects of Scotland.

RIBA: Royal Institute of British Architects.

Rococo: the latter phase of the BAROQUE style, highly ornamental yet more delicate than earlier phases.

Romanesque: revival, from the 6th century onwards, of Roman Imperial culture characterized by use of round ARCH and BASILICA plan.

Romanticism: 18th century sensibility introducing crucial change in how architecture was perceived and therefore formed.

Roof Light: a fixed or opening window in a roof.

Rotunda: a circular or polygonal building which is usually capped with a dome.

Rustication: masonry with a roughened surface and joints marked by deep grooves.

Secessionstil: term to describe the ART NOUVEAU-influenced work of the Secession group, established in Vienna in 1897 by avant-garde artists and architects who had seceded from the conservative *Künstlerhaus*.

Section: a drawing showing a vertical "cut" through a building.

Services: a term referring to the distribution of all utilities, electrical, gas, heating, hot water, air conditioning ducts, telephone cables, etc, throughout a building.

WHO BUILT THAT?

95

Shafted: a pillar is shafted when it has several thin columns attached.

Shingle Style: late 19th century US underivative free style of domestic architecture typified by the hanging of wooden tiles (shingles) on walls.

Sill: the horizontal member at the bottom of a door or window frame. Sometimes "cill".

Skin: the outer envelope or surface membrane of a building, brick walls or the glass and steel CLADDING.

Spandrel: the surface between two ARCHES or RIBS in a vault.

Steel Frame: constructional method based on a calculated framework of steel, BEAMS, girders, COLUMNS and ties.

Strapwork: decorative masonry or plaster in the form of interlaced bands; originated in France and the Netherlands and common in Elizabethan England.

String Course: a moulding or projecting COURSE of stone or brick running horizontally along the face of a building.

Stucco: smooth plaster or cement RENDERING to walls, or moulded for ceilings.

Stupa: a simple Buddhist religious monument often containing relics.

Tecton: team founded in London in 1932 by Berthold Lubetkin and others. The most important representative of the INTERNATIONAL STYLE in the UK, the group was disbanded in 1948.

Timber Frame: method of construction which is based around a frame of interlocking or connected timber beams.

Trabeated: building constructed on the POST-AND-LINTEL principle.

Tracery: ornamental wood or stonework usually found decorating windows.

Transept: the cross-piece of a CRUCIFORM church at right angles to the body or NAVE.

Triforium: a straight, arcaded wall passage facing the NAVE in a church.

Trompe l'œil: wall or ceiling painting which deceives the eye into believing objects portrayed are three-dimensional.

Truss: a number of members framed together to bridge a space.

Tuscan Order: see ORDERS.

UIA: see IUA.

Vault: an arched roof or ceiling of stone, brick or wood designed in a number of shapes. The simplest is the *tunnel* or *barrel vault* designed as a continuous semi-circular archway; the *groin vault*, or *cross vault*, is formed by crossing two tunnel vaults; the *domical vault* is a dome consisting of a number of wedge-shaped sections rather like an umbrella; the *rib vault* is a complex structure formed by arched ribs, and the *fan vault* is a geometric weaving of ribs resembling the structure of a fan. The *stalactite vault* (muqarna) is common in Islamic architecture.

Vernacular: refers to indigenous or traditional building styles.

Volute: the scroll or spiral decoration featured principally in IONIC capitals.

Voussoir: bricks or stones cut into wedge shapes and placed side by side to form an ARCH.

Ziggurat: a stepped pyramidal structure, with a large base receding to a small top.

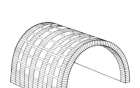
tunnel (barrel) vault

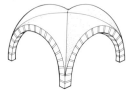
groin (cross) vault

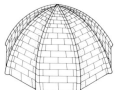
domical vault

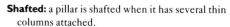
PICTURE CREDITS
Key: *a* = above; *b* = below; *l* = left; *r* = right

p2: Peter Wilson; *p6:* J. Baker Collection; *p7:* J. Baker Collection; *p8:* A. F. Kersting; *p9 (l)* reproduced by kind permission of His Grace the Duke of Marlborough; *p9 (r)* New York Convention and Visitor Bureau; *p10:* C. M. Dixon; *p11:* J. Allan Cash; *p12 (a):* Flora Torrance/Life File; *p12 (b):* Maggie Fagan/Life File; *p13 (a, b):* C. M. Dixon; *p14:* Andrew Ward/Life File; *p15:* J. Allan Cash; *p16:* C. M. Dixon; *p18:* J. Baker Collection; *p19:* J. Baker Collection; *p20:* Eric Wilkins; *p21:* De Cet; *p22:* J. Allan Cash; *p23:* A. F. Kersting; *p24:* A. Gamiet/TRIP; *p25:* David Heath/Life File; *p26:* Peter Wilson; *p27:* Beijing Slide Studio; *p28:* Andrew Watson/Life File; *p29:* Andrew Eapnett/Architectural Association; *p30:* Gina Green/Life File; *p31 (t):* A. F. Kersting; *p31 (b):* B. Harding/TRIP; *p32:* Cliff Threadgold/Life File; *p33:* The Photo Source Library; *p34:* Bob Turner/TRIP; *p35:* J. Baker Collection; *p36:* De Cet; *p37:* A. F. Kersting; *p38:* A. F. Kersting; *p39:* Angelo Hornak Library; *p40:* Andrew Ward/Life File; *p41 (t, b):* Angelo Hornak Library; *p42:* J. Baker Collection; *p43:* J. Baker Collection; *p44:* T. Waeland/Life File; *p45:* J. Baker Collection; *p46:* Colin Penn/Architectural Association; *p47:* J. Allan Cash; *p48:* J. Baker Collection; *p49:* J. Baker Collection; *p50:* British Tourist Authority; *p51:* J. Baker Collection; *p52:* A. F. Kersting; *p53 (t):* Mark Fiennes/Arcaid; *p53 (b):* A. F. Kersting; *p54:* Emma Lee/Life File; *p55 (t, b):* Andrew Ward/Life File; *p56:* Andrew Ward/Life File; *p57:* reproduced by kind permission of His Grace the Duke of Marlborough; *p58:* J. Allan Cash; *p60:* J. Baker Collection; *p61:* Studio Ethel; *p62:* J. Baker Collection; *p63:* Valerie Bennett/Architectural Association; *p64:* Niall Clutton/Arcaid; *p65:* Xavier Catalan/Life File; *p66:* J. Baker Collection; *p67:* De Cet; *p68:* Glasgow School of Art; *p69:* Austrian National Tourist Office; *p70:* Nathan Willock/Architectural Association; *p71:* Scott Frances/Arcaid; *p72 (l):* Bill Chaitkin/Architectural Association; *p72 (r):* J. Baker Collection; *p73:* Dennis Gilbert/Arcaid; *p74:* J. Allan Cash; *p75 (b):* Angelo Hornak Library; *p75 (t):* Julie Waterlow/TRIP; *p76 (t):* New York Convention and Tourist Bureau; *p76 (b):* Richard Bryant/Arcaid; *p77:* J. Allan Cash; *p78:* John Winter/Architectural Association; *p79:* Ezra Stoller/Arcaid; *p80:* D. Tineno/Architectural Association; *p81:* J. Stirling/Architectural Association; *p82:* Australia House Photo Library; *p83:* Life File; *p84:* Richard Bryant/Arcaid; *p85:* Ian Lambot/Arcaid; *p86:* R. Venturi/Architectural Association; *p87:* TRIP; *p88:* Mike Evans/Life File; *p89 (t):* Richard Bryant/Arcaid; *p89 (b):* Peter Mauss Esto/Arcaid; *p90 (t):* Sir Norman Foster and Partners; *p91:* Richard Bryant/Arcaid.

Whilst every effort has been made to trace and acknowledge all copyright holders, we would like to apologize should any omissions have been made.

Title page: The Court of Lions, Alhambra Palace, Granada.